JUST ENOUGH ELECTRONICS TO IMPRESS YOUR FRIENDS AND COLLEAGUES

Bob Zeidman

Cupertino, California

Copyright © 2013 by Bob Zeidman

All rights reserved. No part of this book may be reproduced in any form or by any electronic or mechanical means, including information storage and retrieval systems, without permission in writing from the publisher, except by a reviewer who may quote brief passages in a review.

First paperback edition 2013

ISBN 978-0970227645

10 9 8 7 6 5 4 3 2 1

Book design by Carrie Zeidman

Printed in China

Just Enough Electronics to Impress Your Friends and Colleagues

Table of Contents

1. Introduction 1
   - 1.1 About This Book 1
   - 1.2 About the Author 2
2. Electricity 5
   - 2.1 What is it? 5
   - 2.2 Batteries 8
   - 2.3 Generators 11
     - 2.3.1 Water 11
     - 2.3.2 Steam 13
     - 2.3.3 Wind 16
     - 2.3.4 Active Solar 17
   - Quiz 1: Electricity 19
3. Linear Devices 23
   - 3.1 Resistors 23
   - 3.2 Capacitors 25
   - 3.3 Inductors 28
   - 3.4 Direct Current and Alternating Current 31
   - Quiz 2: Linear Devices 33
4. Electricity as Sound 35
   - 4.1 Electromagnets 35
   - 4.2 Transducers 36
   - 4.3 Microphones and Speakers 36
   - 4.4 Filters 38
   - Quiz 3: Electricity as Sound 43
5. Nonlinear Devices 47
   - 5.1 Diodes 47
   - 5.2 Transistors 49
   - Quiz 4: Nonlinear Devices 53
6. Transmission Lines 57
   - Quiz 5: Transmission Lines 63

7. Digital Logic 65

- 7.1 Boolean Algebra 65
- 7.2 Binary Numbers 68
- 7.3 Schematic Representations 69
- 7.4 Delay Elements 70
- 7.5 Finite State Machines 71
- Quiz 6: Digital Logic 75

8. Semiconductor Technology 79

- 8.1 Conductors, Insulators, and Semiconductors 79
- 8.2 Diodes and pn Junctions 82
- 8.3 Transistors and npn and pnp Junctions 83
- 8.4 Integrated Circuits 84
- 8.5 Growing Silicon 86
- 8.6 Oxidation 87
- 8.7 Lithography and Photoresist 88
- 8.8 Ion implantation and Diffusion 88
- 8.9 Etching 88
- Quiz 7: Semiconductors 91

9. Memory Devices 95

- 9.1 Read Only Memories (ROMs) 95
- 9.2 Nonvolatile Programmable Memories 96
  - 9.2.1 Programmable Read Only Memories (PROMs) 97
  - 9.2.2 Reprogrammable Read Only Memories (EPROMs, EEPROMs, Flash, and NVRAMs) 98
- 9.3 Dynamic Random Access Memories (DRAMs) 101
- 9.4 Static Random Access Memories (SRAMs) 104
- 9.5 Memory Organization 105
- Quiz 8: Memories 107

10. Application Specific Integrated Circuits (ASICs) 109

- 10.1 Gate Array vs. Standard Cell vs. Structured ASIC 110
  - 10.1.1 The Gate Array 110
  - 10.1.2 The Standard Cell 112
  - 10.1.3 The Structured ASIC 114
- 10.2 Gate Array vs. Standard Cell vs. Structured ASIC 116
- Quiz 9: ASICs 117

11. Programmable Devices 121

- 11.1 Programmable Read Only Memories (PROMs) 121
- 11.2 Programmable Logic Arrays (PLAs) 122
- 11.3 Programmable Array Logic (PALs) 123
- 11.4 CPLDs and FPGAs 124
- 11.5 Complex Programmable Logic Devices (CPLDs) 124
  - 11.5.1 CPLD Architectures 125
- 11.6 Field Programmable Gate Arrays (FPGAs) 127
  - 11.6.1 FPGA Architectures 128

- Quiz 10: Programmable Devices 135

12. Computer Architecture 139

- 12.1 System Architecture 139
- 12.2 Processor Architecture 141
  - 12.2.1 Address Register 141
  - 12.2.2 Program Counter 141
  - 12.2.3 General Registers 141
  - 12.2.4 Arithmetic Logic Unit (ALU) 142
  - 12.2.5 Accumulator 142
  - 12.2.6 Instruction Register 142
  - 12.2.7 Control Logic 142
- 12.3 Pipelined Processors 143
- 12.4 Cache Memory 145
- 12.5 RISC vs. CISC 146
- 12.6 Very Long Instruction Word (VLIW) Computers 147
- Quiz 11: Processors 149

13. Engineering Equipment 153

- 13.1 Voltmeter 153
- 13.2 Ohmmeter 154
- 13.3 Ammeter 154
- 13.4 Oscilloscope 155
- 13.5 Signal Generator 157
- 13.6 Logic Probe 158
- 13.7 Logic Analyzer 159
- 13.8 Protocol Analyzer 160
- 13.9 In-Circuit Emulator 161
- Quiz 12: Equipment 163

Just Enough Electronics to Impress Your Friends and Colleagues

14.	**Resources**	**165**
15.	**Index**	**167**
16.	**Answers to Quizzes**	**171**
	Quiz 1: Electricity	173
	Quiz 2: Linear Devices	175
	Quiz 3: Electricity as Sound	177
	Quiz 4: Nonlinear Devices	179
	Quiz 5: Transmission Lines	183
	Quiz 6: Digital Logic	185
	Quiz 7: Semiconductors	189
	Quiz 8: Memories	191
	Quiz 9: ASICs	193
	Quiz 10: Programmable Devices	197
	Quiz 11: Processors	199
	Quiz 12: Equipment	201

1. Introduction

Thank you for purchasing my book. Now that you've bought it, I'll tell you what it's about. And I'll introduce myself.

1.1 About This Book

If you're not an electrical engineer, but you work with them or work for a high tech company, you have a need to understand certain basic concepts of electronics. You might be a technician assisting engineers create the next great society-transforming online service. You might be a marketing guru creating advertising campaigns for the latest communication device. Maybe you're a sales person convincing a large corporation to design your company's quadratic amplifier into their chronosynclastic infundibulum. You might be a patent lawyer. Or you might be a non-engineer managing your company's engineering department. This book will take you on a whirlwind tour of the fundamentals. By the end, you'll be familiar with the basic concepts of various kinds of electronic technologies and devices. You'll be able to impress friends at cocktail parties, enhance your standing at work, and maybe even communicate with the brainy men and women who are changing the world.

This book is based on my popular seminar called *Electrical Engineering for Non-Electrical Engineers* that I've given at conferences around the world. Over the years I've used the feedback from students to try to make it the best electronics overview available. I hope I've succeeded, and I hope I've succeeded in turning that seminar into a useful book, presenting everything a non-electrical engineer might want to know about electronics. At least it attempts to do so by introducing the basic concepts and building on them. You will probably not be able to design computers or computer chips when you're done, but you will understand the basic concepts of how they work and what goes into designing them.

Like any subject, no matter how complex, by understanding

the basics and building upon that understanding, it's possible to understand very complex concepts. The following topics are covered in this book. These topics represent a broad, but critical overview of electronics that builds from simple, basic elements to more complex ones. I believe that these topics form the basis of the majority of electronics projects being pursued today.

- Electricity
- Linear Devices
- Electricity as Sound
- Nonlinear Devices
- Transmission Lines
- Digital Logic
- Semiconductor Technology
- Memory Devices
- Application Specific Integrated Circuits (ASICs)
- Programmable Devices (CPLDs and FPGAs)
- Computer Architecture
- Engineering Equipment

1.2 About the Author

Bob Zeidman is the president and founder of Zeidman Consulting, a premier contract research and development firm in Silicon Valley. Since 1983, Bob has designed computer chips, circuit boards, and electronic systems for laser printers, network switches and routers, hardware emulators, and other complex systems. His clients have included Apple Computer, Cisco Systems, Ricoh Systems, and Stanford University. Bob is also the president and founder of Software Analysis and Forensic Engineering Corporation, the leading provider of software intellectual property analysis tools. Bob is considered a pioneer in the fields of analyzing and synthesizing software source code, having created the

SynthOS™ program for synthesizing operating systems and the CodeSuite® program for detecting software intellectual property theft. Bob is also one of the leading experts in the Verilog hardware description language as well as ASIC and FPGA design.

Bob is a prolific writer and instructor, giving seminars at conferences around the world. Among his publications are numerous articles on engineering and business as well as four textbooks, *The Software IP Detective's Handbook*, *Designing with FPGAs and CPLDs*, *Verilog Designer's Library*, and *Introduction to Verilog*. Bob holds numerous patents and earned two bachelor's degrees, in physics and electrical engineering, from Cornell University and a master's degree in electrical engineering from Stanford University.

Introduction

2. Electricity

For many of you, this section may seem too introductory. If so, just skip it and you won't hurt my feelings. But I need to start somewhere and in writing the other sections I thought, "What if someone really doesn't know the basics of electricity?" The thought kept nagging at me and I hate to make assumptions about people's knowledge. So it seemed to be a good idea to explain electricity and where it comes from.

2.1 What is it?

You probably know what an atom is. It's the fundamental building block of all matter that we know of. Atoms, in turn are made up of smaller particles, but every stable kind of matter (i.e., matter that will not come apart or combine with other matter without some external force) is made up of atoms. The simplest atom known is that of Hydrogen, pictured in Figure 1 while the more complex Carbon atom is shown in Figure 2. Atoms are made up of combinations of three basic particles: neutrons, protons, and electrons. Neutrons and protons are together in the nucleus while electrons fly around them. Although the electrons are often drawn orbiting the nucleus like planets orbiting the sun, we now understand that the electrons don't make nice, simple, circular trajectories. Nonetheless it's easy to picture them that way. Protons and electrons attract each other with a force called the electromagnetic force. Protons repel other protons and electrons repel other electrons with the same electric force. Protons are considered to have a positive electric change while electrons are considered to have a negative electric charge.

Electricity

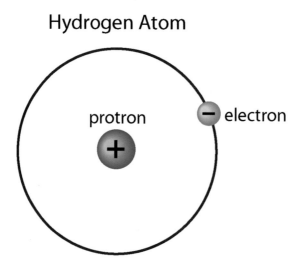

Figure 1. An atom of Hydrogen

Electricity is simply a flow of electrons. Normally electrons want to stay close to their counterpart protons in an atom, but certain kinds of atoms, called conductors, have some electrons that are so far out from the nucleus that a little energy will cause them to pop over to a nearby atom. This pops an electron off that nearby atom that then jumps over to another nearby atom. This chain reaction of electrons jumping from one atom to a nearby atom is called electric current, measured in amps, named after the French mathematician and physicist André Marie Ampère who developed the fundamental theory of electricity and magnetism.

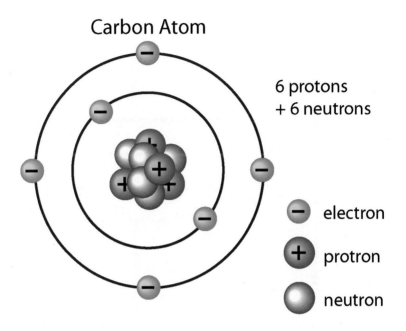

Figure 2. An atom of Carbon

Metals are excellent conductors. We can create a long, thin wire and place a source of electrons at one end that are energized so that they want to get away from each other. For example, if we force a large number of electrons together, because they repel each other, they will attempt to move away. If we attach a copper wire to this large group of electrons, one electron will jump to one of the copper atoms, knocking one of the copper atom electrons off to the next copper atom. As long as we can keep a large enough group of electrons at one end of the wire, we will be able to keep this electric current of electrons jumping down the length of the wire to the other end. This basic principle of electrical engineering was discovered by the great Benjamin Franklin, whose many discoveries and accomplishments in just about all aspects of human endeavor are truly awe inspiring.

2.2 Batteries

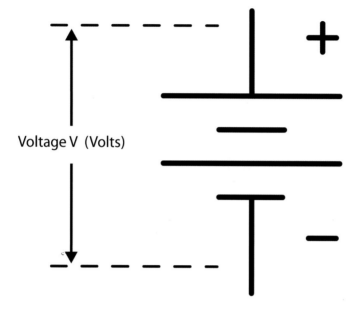

Figure 3. The symbol for a battery

As I mentioned above, in order to get electrons flowing, we need to have a source of electrons. One source of electrons is the battery, the symbol for which is shown in Figure 3, consisting of one or more electrochemical cells, invented in 1792 by Alessandro Volta. These electrochemical cells use a chemical reaction to separate electrons from their atoms. These cells, also known as voltaic cells after their inventor, have a chemical reaction going on inside them. Inside the battery are two different kinds of electrodes that form the terminals of the battery. The electrodes are separated but surrounded by a substance called an electrolyte. One electrode, called the anode, is composed of a material that reacts with the electrolyte to produce positive ions, which are molecules of the electrolyte with electrons have gotten knocked off of them. The other electrode, called the cathode, is composed of a material that reacts with the electrolyte to produce negative ions, which are molecules of the electrolyte that snags the extra electrons. So overall, the electrons have shifted from one end of the battery to the other. There are no more electrons in the battery than before, but now they are grouped at one end where they don't like being around each other.

Throughout this section I will use the analogy of water and pipes to think of electricity and electrical circuits. So a battery is like a bucket of water held up high like the one shown in Figure 4. The water has the potential to fall down but needs something, like a tip of the bucket, to get it started. In the same way, a battery has the potential to send electricity around a circuit, sending the electrons from the negative terminal where there are too many electrons back to where they want to be— at the positive terminal where there aren't many electrons. But it requires something else to get that electric current moving. In the case of the bucket of water, it requires a slight tilt to give the water a path to the ground. With the battery it requires a wire to give the electrons a path from the cathode of the battery to the anode. This potential for electric current, the "electric potential" is measured in volts. The more volts a battery has ("higher voltage"), the more potential it has to send electrons around a circuit[1].

---

[1] One thing that used to confuse me is that while electrons physically travel from the negative terminal to the positive terminal of a battery, electric current is defined as going from the positive terminal to the negative one. That's because Benjamin Franklin discovered electricity but didn't know about atomic particles like electrons. He knew that electricity traveled around a circuit, but didn't know which direction. For practical purposes it didn't really matter so he just picked one direction and called it positive current. When atomic particles were discovered, years later, it was determined that it was negatively charged electrons that were physically moving. So we're now stuck with the concept of positive electrical current in one direction really being the movement of negatively charged electrons in the opposite direction.

Electricity

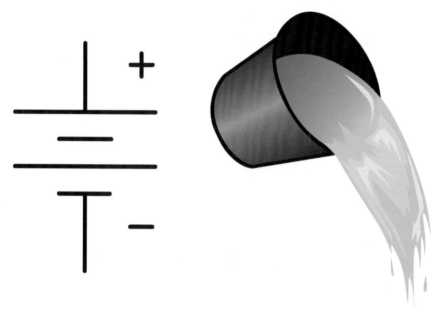

Figure 4. Battery as a water bucket

Though most people have seen lots or batteries, a picture of different kinds of batteries is shown in Figure 5.

Figure 5. Batteries

2.3 Generators

Another method of creating electric current is by using what is called an electric generator. These generators all work according to the same principle, which involves moving a coil of wire between the poles of a magnet or, conversely, moving magnets around a coil of wire. Magnetism and electricity are different forms of the same force, electromagnetism. Electricity is the force due to electrons and protons that are essentially stationary. When they are moved around, the properties of the force appear different and the force is called magnetism. In a permanent magnet, like the one you might attach to your refrigerator door, there are very small electric currents inside the material that are causing the magnetic effect.

When a metal coil is moved within a magnetic field, it causes the electrons to jump off their atoms, resulting in an electric current in the coil. These moving parts are incorporated into turbines. That's the easy part. The difficult part is getting the turbines moving, and that's where different solutions have been found over the years. Some of the solutions include the following.

2.3.1 Water

Flowing water turns turbine blades. The water can be harnessed from the power of a waterfall such as Niagara Falls shown in Figure 6 or in a dam, such as the Hoover Dam shown in Figure 7, two famous hydroelectric power generation plants.

Electricity

Figure 6. Niagara Falls

Figure 7. Hoover Dam

2.3.2 Steam

Steam is by far the most commonly used method to turn turbines to generate electricity, but there are many ways to generate steam including geothermal power, fossil fuels, passive solar power, and nuclear power.

2.3.2.1 Geothermal

The geothermal method of creating steam to power turbines to generate electricity involves tapping into geysers where spring water is heated naturally under the ground, turns into steam at high pressure, spinning a turbine to generate electricity, as shown in the geothermal power plant of Figure 8.

2.3.2.2 Fossil fuels

Coal, natural gas, and petroleum are burned to heat water into steam that in turn spins the turbine that generates electricity as shown in the power plant of Figure 9.

Electricity

Figure 8. Geothermal power plant

Figure 9. Fossil fuel power plant

2.3.2.3 Passive solar

Heat from the sun is concentrated, typically using mirrors, to heat up water, turning it into steam to spin the turbine to generate electricity as shown in the passive solar power plant of Figure 10.

Figure 10. Passive solar power plant

2.3.2.4 Nuclear

The heat generated from a nuclear fission reaction (splitting atoms) is used to heat water to spin turbines to generate electricity as shown in the nuclear power plant of Figure 11.

Figure 11. Nuclear power plant

2.3.3 Wind

Wind can be used to spin turbines to generate electricity such as those built in 1887 by Scottish professor James Blyth. These power windmills supply a significant percentage of Denmark's power and are landmarks of Northern California at the Altamont Pass shown in Figure 12.

Figure 12. Windmills at Altamont Pass

2.3.4 Active Solar

Although solar can be used to heat water into steam and turn turbines in the more traditional way of generating power, solar can also be used to more directly generate electricity using photovoltaic panels of solar cells made from semiconductors as shown in Figure 13. As we will learn later in the book, semiconductors are materials whose atoms don't like to give up their electrons and so they are not good at conducting electricity, except if energy is applied and then their electrons happily bounce off to other atoms, turning into conductors. In these cells, sunlight provides the energy to excite the electrons into moving faster and eventually bouncing around and creating an electric current. This is called the photovoltaic effect. Its older brother the photoelectric effect was explained by Albert Einstein in 1905 and it was this discovery, not the Theory of Relativity for which he is popularly known, that won him the Nobel Prize in Physics in 1921.

Figure 13. Solar cells

Just Enough Electronics to Impress Your Friends and Colleagues

Quiz 1: Electricity

1. **Electricity is**
   - ☐ the motion of atoms
   - ☐ the flow of electrons
   - ☐ the chemistry between two people
   - ☐ the potential for the flow of electrons

2. **Which of the following have a positive electric charge?**
   - ☐ electron ☐ proton ☐ neutron ☐ nucleus

3. **Which of the following have a negative electric charge?**
   - ☐ electron ☐ proton ☐ neutron ☐ nucleus

4. **Which of the following have no electric charge?**
   - ☐ electron ☐ proton ☐ neutron ☐ nucleus

5. **A chain reaction of electrons jumping from one atom to a nearby atom is called**
   - ☐ a Conga line
   - ☐ electricity
   - ☐ a nuclear reaction
   - ☐ the photoelectric effect

6. **What uses a chemical reaction to separate electrons from their atoms?**
   - ☐ electrochemical cells
   - ☐ voltaic cells
   - ☐ chemoelectric cells
   - ☐ the photoelectric effect

7. **A battery consists of one or more**
   - ☐ electrochemical cells
   - ☐ voltaic cells
   - ☐ chemoelectric cells
   - ☐ the photoelectric effect

Electricity

8. **Batteries produce electricity by**
   - ☐ moving a coil of wire between the poles of a magnet
   - ☐ separated electrodes surrounded by an electrolyte
   - ☐ moving magnets around a coil of wire
   - ☐ the photoelectric effect

9. **Generators produce electricity by**
   - ☐ moving a coil of wire between the poles of a magnet
   - ☐ separated electrodes surrounded by an electrolyte
   - ☐ moving magnets around a coil of wire
   - ☐ the photoelectric effect

10. **Which of the following types of generators do not heat water into steam**
    - ☐ passive solar
    - ☐ geothermal
    - ☐ nuclear
    - ☐ water
    - ☐ active solar

11. **Throughout this book, to describe electric circuits I will use the analogy of**
    - ☐ Conga lines
    - ☐ Newton's Cradle
    - ☐ lasers and mirrors
    - ☐ water moving through pipes

12. **Albert Einstein was awarded his Nobel Prize for**
    - ☐ the special theory of relativity
    - ☐ the photovoltaic effect
    - ☐ his work against the atomic bomb
    - ☐ the general theory of relativity

Electricity

3. Linear Devices

The very basic components of electronics are the resistor, capacitor, and inductor. These components control the flow of electricity through a circuit, which is really the key to designing all electronic circuits. Essentially, by controlling electric currents through a circuit, the engineer is able to use that circuit to perform real functions such as opening a garage door, calculating an arithmetic expression, or showing a picture on a TV screen. Once the electricity can be controlled, it can be used to represent things from numbers in a computer calculation to logical decisions in a controlling device to videos streaming from a website.

These devices are called linear components because they control the flow of electricity in a linear way. In other words, the operations they perform can be expressed using only multiplication and division.

3.1 Resistors

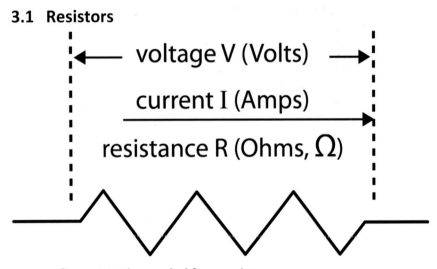

Figure 14. The symbol for a resistor

Resistors, simply put, resist electric current going through them. Figure 14 shows the symbol for a resistor, the jagged line. There is a voltage V, measured in volts, across the terminals of the resistor that represents the potential of the electricity. You can think of voltage as how strongly the electrons would like to get across that resistor. There is also an electric current *I*, measured in amps, that represents how much electricity is actually getting across the resistor. The ratio of the voltage to the current is the resistance R, a property of the resistor measured in ohms, named after Georg Ohm, the German physicist who discovered this relationship. Put another way, if you know the voltage and the resistance, the current will be determined by Equation 1 known as Ohm's Law and is the basis of all electronics.

$$I = V/R$$

Equation 1. Ohm's Law

A resistor can be thought of like a restriction in a water pipe, which reduces the flow of water through the pipe. Similarly, a resistor restricts the flow of electrons through the wire. This water pipe model is shown in Figure 15.

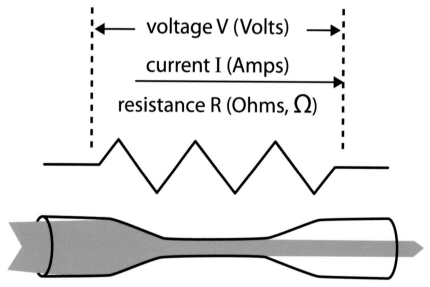

Figure 15. Resistor as water pipe

Examples of resistors can be seen in the picture in Figure 16.

Just Enough Electronics to Impress Your Friends and Colleagues

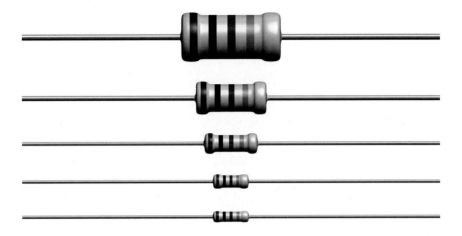

Figure 16. Resistors

3.2 Capacitors

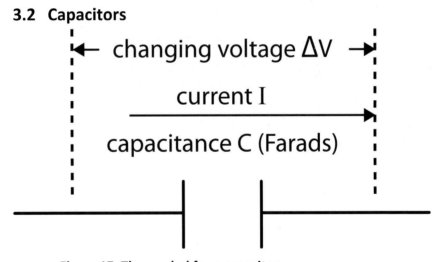

Figure 17. The symbol for a capacitor

Capacitors store electric charge. Figure 17 shows the symbol for a capacitor, two terminals on either side of two parallel lines. As with a resistor, there is a voltage V, measured in volts, across the terminals of the capacitor that represents the potential of the electricity. There is also an electric current I, measured in amps, that represents how much electricity is actually getting across the capacitor. Unlike a resistor, current will only go across the capacitor long enough to fill it up and then it stops. But if you keep changing the voltage across it, the capacitor will either allow more current across and capture more electric charge (by increasing the voltage) or spill electric charge out and push current backwards (by decreasing the voltage). The mathematical symbol for change is the delta (triangle symbol). The relationship between changes in the current and changes in the voltage over time across a capacitor depends on how much electric charge the capacitor can store. The storage ability of a capacitor is called its "capacitance" and is measured in farads, named after British chemist and physicist Michael Faraday who performed some of the first and most fundamental experiments in electricity and magnetism. The equation that shows how these quantities depend on each other is given in Equation 2.

$$I = C(\Delta V/\Delta t)$$

Equation 2. Capacitor equation

In this equation, I is the current, ΔV is the change in the voltage, and Δt is the amount of time it takes the voltage to change (Δ is pronounced "delta"). C is the capacitance, a property of the capacitor measured in farads. This equation shows how much electric current goes through the circuit as the voltage is changed over time. Multiply the capacitance by the change in voltage divided by the amount of time it takes to change, and you will find how much current is generated.

If resistors can be thought of as restrictions in a water pipe, capacitors can be thought of as buckets that hold the water, as shown in Figure 18.

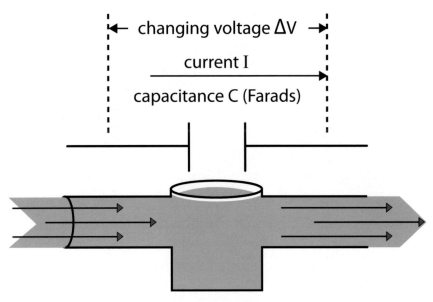

Figure 18. Capacitor as water bucket

One significant difference between resistors and capacitors is that resistors affect current and voltage in a constant way. The effect of capacitors on current and voltage depend on changes to the voltage and current. If someone tells you what the current is through a resistor, you can use Ohm's Law to tell you what the voltage must be. If someone tells you what the voltage is across a capacitor, you must ask how much did it change over the last second or microsecond or nanosecond in order to determine how much current went through it. This property is very important for manipulating electrical signals.

Examples of capacitors can be seen in the picture in Figure 19.

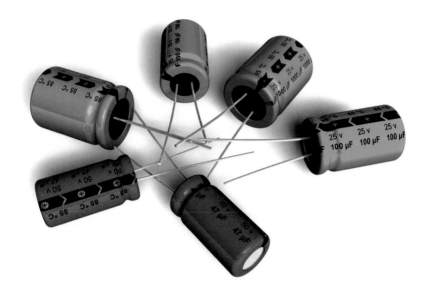

Figure 19. Capacitors

3.3 Inductors

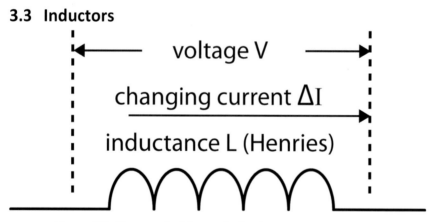

Figure 20. The symbol for an inductor

Figure 20 shows the symbol for an inductor, two terminals on either side of a stretched coil. Inductors, like capacitors, depend on changes in voltage and current. There is a voltage *V* across the terminals of the capacitor that represents the potential of the electricity. There is also a changing electric current *I* that represents how much electricity is actually getting across the inductor.

When a voltage is applied across an inductor, the current continually increases over time. The relationship between the voltage at any particular time and changes in the current across an inductor over time depends on the inductor's "inductance," measured in henries, named after American Joseph Henry who was one of the first electrical engineers, inventing devices that utilized electric and magnetic forces. The equation that shows how these quantities depend on each other is given in Equation 3.

$$V = L(\Delta I/\Delta t)$$

Equation 3. Inductor equation

In this equation L is the inductance, V is the input voltage, ΔI is the change in the current, and Δt is the amount of time it takes the current to change. Multiply the inductance by the change in current divided by the amount of time it takes to change, and you will find how much voltage was applied to the inductor.

Inductors can be thought of as waterwheels in the pipe, because they are difficult to get started, and you must push a lot of water through the pipe to get it going and get water out the other end. Once they get going, though, the inertia keeps them going and they don't want to stop. This can be seen in Figure 21.

Linear Devices

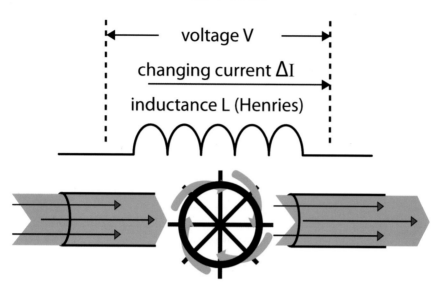

Figure 21. Inductor as a water wheel

Examples of inductors can be seen in the picture in Figure 22.

Figure 22. Inductors

3.4 Direct Current and Alternating Current

The kind of electricity we've been talking about so far is called direct current or DC. This means that current travels in one direction directly from a beginning terminal (typically the positive anode of a battery), around a circuit, to an ending terminal (typically back to the negative cathode of a battery). Direct current is simple to understand and was the kind of electric power championed by one of the greatest inventors in history, Thomas Edison.

Alternating current or AC involves electrons that periodically switch directions. This switching direction causes surges in voltage and current at any point in a circuit that are described by a sine wave such as the one in Figure 23.

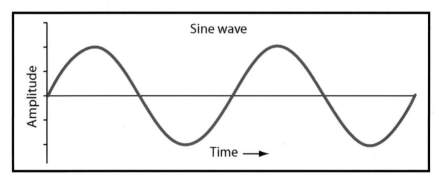

Figure 23. Alternating current sine wave

Alternating current or AC was perfected by Nicola Tesla, a mathematical and electrical genius who left Edison's employment to join entrepreneur George Westinghouse to compete against his former boss. AC is more difficult to generate and requires advanced mathematics to understand, but it allows for more efficient distribution over wide networks, which is why it won out as the kind of electricity supplied to homes and businesses via electrical grids throughout the world. In America, electricity supplied to houses is 120 volts and 60 cycles per second or Hertz, named for Heinrich Hertz, the German physicist who proved that electromagnetic waves

Linear Devices

existed. This means that the voltage at any one spot averages 120 volts, but is changing 60 times per second. In Europe, the average voltage varies between 220 and 240 volts, depending on the country, changing at 50 hertz.

Just Enough Electronics to Impress Your Friends and Colleagues

Quiz 2: Linear Devices

1. Match the component name in the box next to the corresponding symbol.

—/\/\/—	Resistor
—| |—	Capacitor
—ᵐᵐ—	Inductor

2. Which of the following is not a basic component of electrical engineering?

    ☐ resistor   ☐ gyrator   ☐ capacitor   ☐ inductor

3. Components whose properties can be described using multiplication or division are called:

    ☐ linear   ☐ nonlinear   ☐ colinear   ☐ planar

4. The equation that states I=V/R is called:

    ☐ Moe's Law   ☐ Gomez's Law

    ☐ Ohm's Law   ☐ Murphy's Law

Linear Devices

5. **A resistor can be thought of as a:**
   ☐ pipe with a water wheel   ☐ pipe with a bucket
   ☐ pipe with a cinch in it   ☐ pipe with tobacco in it

6. **Resistance is measured in:**
   ☐ farads   ☐ ohms   ☐ amps   ☐ henries

7. **Current is measured in:**
   ☐ watts   ☐ ohm   ☐ amps   ☐ henries

8. **Capacitance is measured in:**
   ☐ farads   ☐ ohms   ☐ amps   ☐ henries

9. **A capacitor can be thought of as a:**
   ☐ pipe with a water wheel   ☐ pipe with a bucket
   ☐ pipe with a cinch in it   ☐ pipe with tobacco in it

10. **Inductance is measured in:**
    ☐ farads   ☐ ohms   ☐ amps   ☐ henries

11. **An inductor can be thought of as a:**
    ☐ pipe with a water wheel   ☐ pipe with a bucket
    ☐ pipe with a cinch in it   ☐ pipe with tobacco in it

12. **Hertz is a measure of:**
    ☐ cycles per second   ☐ direct current
    ☐ wavelength   ☐ seconds per cycle

4. Electricity as Sound

One of the most important inventions of the twentieth century was the telephone by Alexander Graham Bell. The primary great conceptual invention of the telephone involved converting sound waves into electricity, transmitting the electricity to another location, and then converting the electricity back to sound waves. To understand how this happens, you need to first understand electromagnets, transducers, microphones, and speakers.

4.1 Electromagnets

Earlier I discussed how magnets could be used to generate electricity. The same is true in reverse. Electricity can be used to create magnets. By wrapping an insulated metal wire around a conductor "core" and passing an electric current through the wire coil, the conductor core can be turned into a magnet as shown in Figure 24. If the core is made of a material called ferromagnetic, the core can be a very powerful magnet that is turned on and off and its poles can even be reversed when the current is turned on and off and reversed respectively.

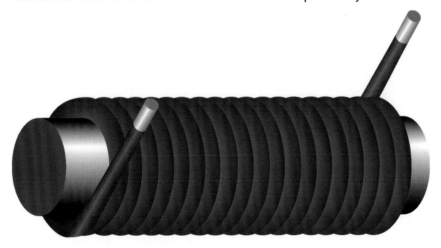

Figure 24. Electromagnet

4.2 Transducers

Transducers convert pressure into electricity. The pressure of a varying sound wave gets turned into a similarly varying current or voltage. A transducer can use magnets. When the magnets are vibrated by a sound wave, they induce an electric current in a wire. Or a transducer can use a capacitor. When the capacitor's plates vibrate, the amount of electric charge it can hold changes and so it either collects more electrons or spits them out. Either of these transducers, or other kinds of transducers, such as those that use piezoelectric generation or light modulation, produce an electrical signal from mechanical vibration. An example of a transducer that uses a moving magnet to produce an electric current in a coil is shown in Figure 25.

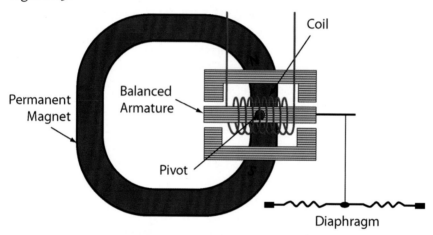

Figure 25. Transducer

4.3 Microphones and Speakers

Microphones use transducers to convert vibrations from sound waves into similarly varying electrical signals. Speakers use electromagnets to move materials, typically paper cones, to convert those varying electrical signals back into sound. This conversion process is so ubiquitous in today's society, and so important to modern communication, that we tend to take this amazing discovery for granted. A cross section of a microphone is shown in Figure 26 while a cross section of a speaker is shown in Figure 27.

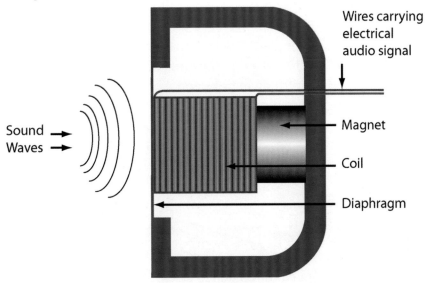

Figure 26. Microphone cross section

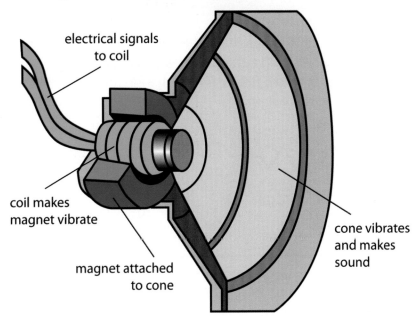

Figure 27. Speaker cross section

The entire process of converting sound to electricity, transmitting it over distances, and converting it back to sound is shown in Figure 28. These electrical signals that are generated from sound waves can change very quickly, called high frequency. Or they can change very slowly and are called low frequency. Most sound waves, and their corresponding electrical signals, contain a complex mix of high frequency, low frequency, and everything in between. For sound, this is known as treble, bass, and mid-range.

Figure 28. Sound conversion to electricity and back to sound

4.4 Filters

The different linear devices that we learned about earlier—resistors, capacitors, and inductors—can be connected in various ways to produce filters, which change the properties of an incoming electrical signal.

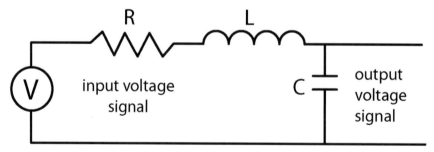

Figure 29. Low pass filter

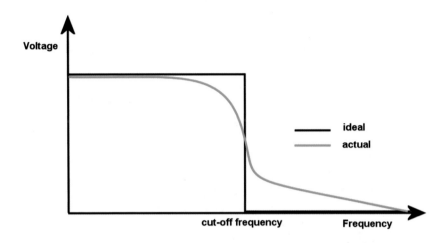

Figure 30. Low pass filter voltage vs. frequency

Figure 29 shows a circuit for a low pass filter where the circle with a V inside represents some voltage source that is varying. For example it could be the electrical signal coming from a microphone. Figure 30 shows a diagram of how that filter affects the frequencies of electrical signals. A low pass filter takes in signals of all frequencies, but only allows the low frequency signals, the slowly changing signals, to pass through. By adjusting the values of the resistor, capacitor, and inductor, you can select the exact cut-off frequency where signals below that frequency are passed through while signals above that frequency are not.

These low pass filters that work on the electrical signals generated by sounds are the bass control on your stereo system. They are also used to reduce the hiss that might come from your tape deck (if you know what a tape deck is).

Electricity as Sound

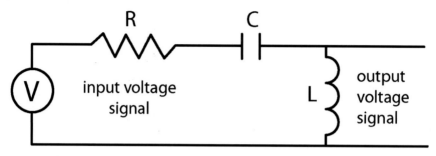

Figure 31. High pass filter

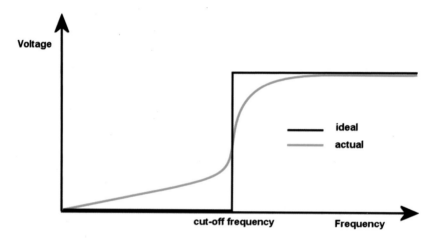

Figure 32. High pass filter voltage vs. frequency

Figure 31 shows a circuit for a high pass filter and Figure 32 shows a diagram of how that filter affects the frequencies of electrical signals. High pass filters only let through the quickly changing, or high frequency signals. By adjusting the values of the resistor, capacitor, and inductor, you can select the exact cut-off frequency where signals above that frequency are passed through while signals below that frequency are not. This is the treble control on your stereo system.

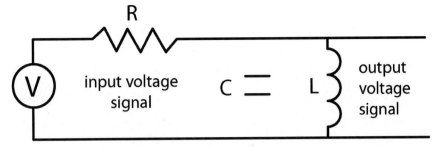

Figure 33. Band pass Filter

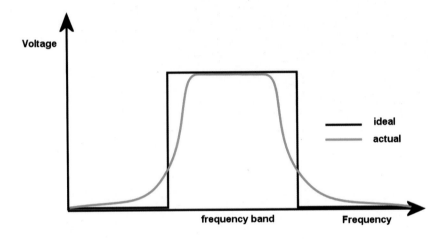

Figure 34. Band pass filter voltage vs. frequency

Figure 33 shows a circuit for a band pass filter and Figure 34 shows a diagram of how that filter affects the frequencies of electrical signals. Band pass filters let through only specific frequencies but don't let higher or lower frequency signals pass. By adjusting the values of the resistor, capacitor, and inductor, you can select the exact frequency band where only signals within a certain range of frequencies are passed through while all other frequencies are not. These are the mid-band controls on a graphic equalizer in your stereo system, like the one in Figure 35.

Figure 35. Graphic equalizer on a stereo system

Quiz 3: Electricity as Sound

1. **An electromagnet is created by**
   - ☐ connecting a magnet to a battery
   - ☐ wrapping a wire around a core and passing an electric current through it
   - ☐ wrapping a wire around a battery
   - ☐ passing an electric current through a wire coil while moving it

2. **The best core for an electromagnet is made from which material?**
   - ☐ electromagnetic
   - ☐ hydrogen
   - ☐ fermium
   - ☐ ferromagnetic

3. **What does a transducer do?**
   - ☐ convert electricity into pressure
   - ☐ transmit changing electrical signals
   - ☐ induce current in wires
   - ☐ convert pressure into electricity

4. **What does a microphone uses a transducer to do?**
   - ☐ convert sound to electricity
   - ☐ a microphone does not use a transducer
   - ☐ filter out different frequencies
   - ☐ convert electricity to sound

Electricity as Sound

5. **What does a speaker uses an electromagnet to do?**
   ☐ convert sound to electricity
   ☐ a speaker does not use an electromagnet
   ☐ filter out different frequencies
   ☐ convert electricity to sound

6. **The frequency of a signal describes**
   ☐ how quickly the signal changes
   ☐ the intensity of the signal
   ☐ the decibel of the signal
   ☐ the mean square root of the voltage of the signal

7. **A low pass filter lets through**
   ☐ high frequency signals   ☐ low frequency signals
   ☐ all signals   ☐ no signals
   ☐ only signals in a certain range of frequencies

8. **A high pass filter lets through:**
   ☐ high frequency signals   ☐ low frequency signals
   ☐ all signals   ☐ no signals
   ☐ only signals in a certain range of frequencies

9. **A band pass filter lets through**
   ☐ high frequency signals   ☐ low frequency signals
   ☐ all signals   ☐ no signals
   ☐ only signals in a certain range of frequencies

10. **The bass control on your stereo is**
    - ☐ a high pass filter
    - ☐ a low pass filter
    - ☐ a band pass filter
    - ☐ turned up way too loud

11. **The treble control on your stereo is**
    - ☐ a high pass filter
    - ☐ a low pass filter
    - ☐ a band pass filter
    - ☐ turned up way too loud

12. **The cut-off frequencies of a filter are determined by**
    - ☐ the value of the resistor
    - ☐ the value of the capacitor
    - ☐ the value of the inductor
    - ☐ all of the above

Electricity as Sound

5. Nonlinear Devices

Nonlinear devices are devices that cannot be defined by a simple multiplication or division equation, because they behave one way under some conditions and another way under other conditions. Diodes and transistors are two of the most basic nonlinear electronic devices.

5.1 Diodes

The simplest nonlinear device is a diode, invented by Sir Ambrose Fleming, an English electrical engineer and physicist. A diode acts like a one-way valve letting electricity flow without restriction in one direction, from the anode terminal to the cathode terminal, while not letting it flow at all in the other direction. This can be seen in Figure 36. If a circuit is set up to get current flowing in the forward direction, the diode essentially acts as a wire letting the electricity flow freely. If a circuit is set up to get current flowing in the backward direction, the diode acts like an insulator and does not let the current pass through.

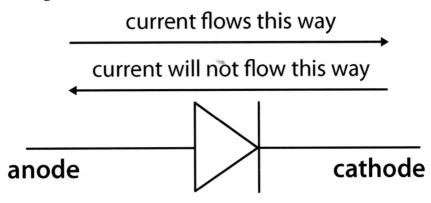

Figure 36. The symbol for a diode

The water flow analogy to a diode is a one-way valve. As shown in Figure 37, a one-way valve lets water flow in one direction but not the other in the same way that a diode allows electric current to flow in one direction but not the other.

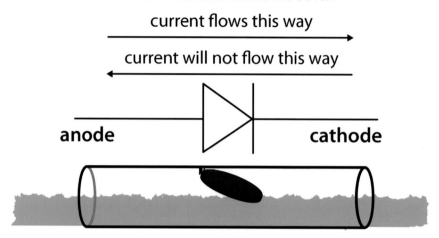

Figure 37. Diode as a one-way valve

Examples of diodes can be seen in the picture in Figure 38.

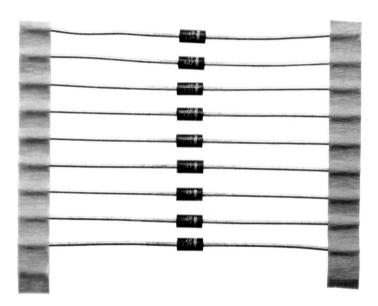

Figure 38. Diodes

5.2 Transistors

Perhaps the most important invention of the twentieth century is the nonlinear device known as the transistor. Its inventors, Walter Brattain, William Shockley, and John Bardeen, created this remarkable device at AT&T Bell Labs, the research arm of American Telephone & Telegraph Company, which grew out of Bell Telephone Company, founded by Alexander Graham Bell to market and sell his telephone many years before. For their efforts, these American inventors were awarded the Nobel Prize in Physics.

The transistor is an electrically controlled switch that allows one circuit to turn another circuit on or off. One basic kind of transistor, the bipolar junction transistor, or BJT, is shown in Figure 39. There are three terminals to a BJT transistor, the collector, the base, and the emitter. Electric current goes into the collector and out of the emitter, but only under the control of the current going into the base. If you turn off the current into the base, then the transistor closes, like a backwards diode, and does not allow any current through it, as if it were an insulator. If you put only a little current into the base, the transistor turns on and allows current from the collector to the emitter as if it were just a wire[2].

---

[2] When I was in my teens I tried to study transistors. Most books on electronics showed that when current was put through the transistor's base, the amount of current going into the transistor's collector came out the transistor's emitter. But I kept asking what happened to the current going into the base? After struggling with this question for a long time, and reading several reference books, I gave up. I believed that there was some phenomena going on that allowed current to go into the device and disappear. It was so disconcerting that I gave up my study of electronics altogether until I got to college. There I discovered that the base current goes out the emitter as shown in my diagram. Since the base current is only a tiny, tiny fraction of the collector-to-emitter current, it is usually ignored. I have shown it here, and described it in this footnote, so that other budding electrical engineers won't be confused and frustrated by this often-unstated approximation.

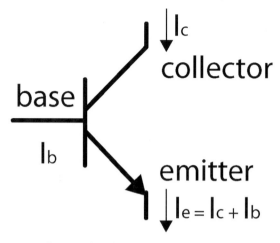

Figure 39. The symbol for a bipolar transistor (BJT)

The other basic kind of transistor is the field effect transistor or FET, shown in Figure 40. The principle of the FET is the same as the BJT, both acting as an electrically controlled switch. The BJT controls the switch using a small current. The FET controls the switch using a small voltage. There are three terminals to an FET transistor, the source, the gate, and the drain. When there is no voltage applied to the gate, the transistor acts like an insulator, preventing current between the source and drain. When a small voltage is applied to the gate, the transistor acts like a wire allowing current from the source to the drain.

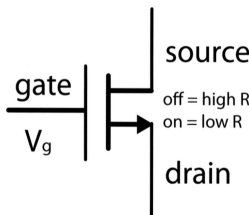

Figure 40. The symbol for a field effect transistor (FET)

Using our water analogy, the transistor acts as a valve controlled by another water pipe, allowing one circuit to turn another circuit on or off as shown in Figure 41. The transistor was very small and used significantly less power than the vacuum tubes it replaced. It revolutionized radio, allowing hand-held radios to be manufactured cheaply that could be powered by inexpensive household batteries (the "transistor radio"). It revolutionized practically every area of electronics. Perhaps most importantly, the transistor's switch effect that allows one electrical circuit to control the behavior of another electrical circuit is the basis of the modern computer that has so greatly changed civilization.

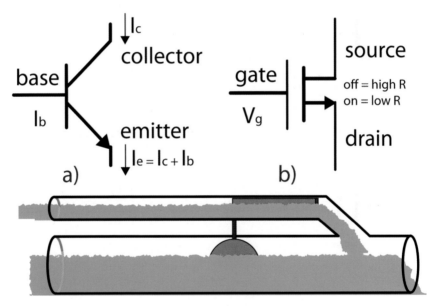

Figure 41. Transistor as a controlled valve

An example of a transistor can be seen in the picture in Figure 42.

Nonlinear Devices

Figure 42. Transistor

Quiz 4: Nonlinear Devices

1. Enter the letter for the component name in the box next to the corresponding symbol.

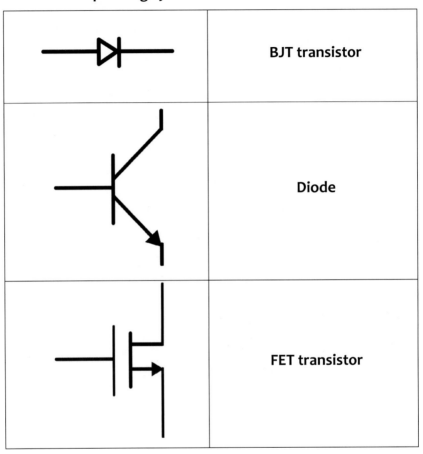

2. Components that cannot be defined by a simple multiplication or division equation are called

    ☐ linear   ☐ nonlinear   ☐ co-linear   ☐ planar

3. Select the terminals of a diode

    ☐ gate   ☐ collector   ☐ cathode   ☐ emitter

    ☐ drain   ☐ base   ☐ source   ☐ anode

Nonlinear Devices

4. **A diode can be thought of as a**
   - ☐ pipe with a valve that is controlled by another pipe
   - ☐ pipe with a bucket
   - ☐ pipe with a cinch in it
   - ☐ pipe with a one-way valve

5. **A transistor can be thought of as a**
   - ☐ pipe with a valve that is controlled by another pipe
   - ☐ pipe with a bucket
   - ☐ pipe with a cinch in it
   - ☐ pipe with a one-way valve

6. **Select the two basic kinds of transistors.**
   - ☐ field effect transistors
   - ☐ bipolar disorder transistors
   - ☐ bipolar junction transistors
   - ☐ Frankenstein effect transistors

7. **Select the terminals of a BJT transistor**
   - ☐ gate   ☐ collector   ☐ cathode   ☐ emitter
   - ☐ drain ☐ base   ☐ source   ☐ anode

8. **In a BJT transistor, the base current controls the flow of electricity from**
   - ☐ the base to the emitter
   - ☐ the emitter to the collector
   - ☐ the source to the drain
   - ☐ the collector to the emitter

9. **Select the terminals of an FET transistor**
   - ☐ gate   ☐ collector   ☐ cathode   ☐ emitter
   - ☐ drain   ☐ base   ☐ source   ☐ anode

10. **In an FET transistor, the gate voltage controls the flow of electricity from**
    - ☐ the gate to the drain
    - ☐ the emitter to the collector
    - ☐ the source to the drain
    - ☐ the collector to the emitter

11. **The invention of transistors enabled**
    - ☐ inexpensive radios
    - ☐ low power radios
    - ☐ modern computers
    - ☐ small, hand-held radios
    - ☐ efficient telephone circuits
    - ☐ all of the above

12. **Transistors replaced**
    - ☐ vacuum tubes
    - ☐ transducers
    - ☐ electromagnets
    - ☐ resistors

Nonlinear Devices

6. Transmission Lines

Transmission lines are essentially any conductor that acts like a combination of many resistors, capacitors, and inductors. When this happens, any change in voltage at one end of the transmission line propagates down to the other end, being absorbed and re-emitted by all of these linear devices. When this voltage wave reaches the end of the transmission line, it is either absorbed or reflected back. If it is reflected back, the absorption and re-emission process continues creating an electrical wave in the opposite direction until it reaches the starting point where it is again either absorbed or reflected back. The process continues until, at some point, equilibrium is reached and each part of the transmission line is at the same voltage as shown in Figure 43.

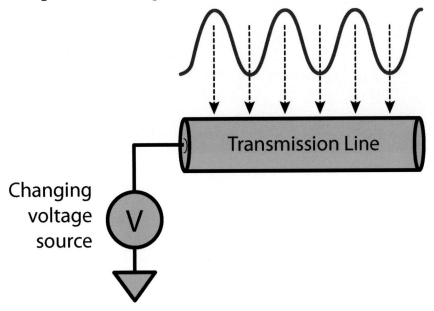

Figure 43. Transmission line

Note that the small triangle in the lower left corner of the diagram represents ground voltage. This is where zero volts is defined for the system. Voltage is actually relative—one point is a higher or lower voltage than another. At some point engineers and scientists decided to call our planet earth zero volts because it can basically take a very large number of electrons and quickly distribute them without any noticeable change in the earth's charge because it's so large. Zero volts is also called "earth ground" or simply "ground." Any voltage greater than ground is considered a positive voltage. Any voltage less than ground is considered a negative voltage.

Figure 44. Power transmission line

Transmission lines are the lines that carry electric power throughout the country, like the one shown in Figure 44. Fine tuning them so that they carry electric current efficiently is an important aspect of power electronics. However, all conductors act as a transmission line to some extent.

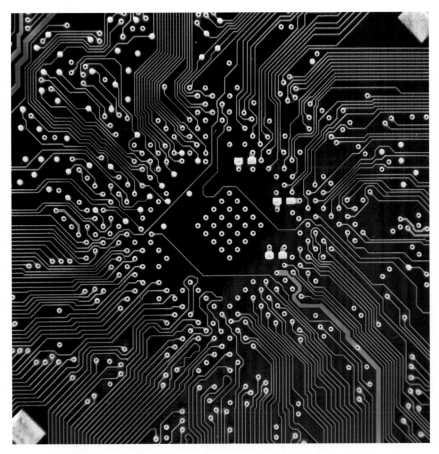

Figure 45. Traces on a printed circuit board

A printed circuit board (PCB) has flat metal strips, typically copper, on the surfaces of the boards and in between layers of the board to carry electric signals. These flat metal strips are called traces, like those shown in Figure 45. When a device like an integrated circuit on a PCB turns on and applies a voltage to a metal trace (called "driving" that trace), that voltage change from zero volts to some other voltage propagates down the trace. When more than one device turns on or off, multiple waves travel through the traces, colliding with each other and either adding to create larger voltages or subtracting to create smaller voltages or doing both at different times. Years ago, electronic devices turned on and off very slowly compared with the time for a wave to propagate along a trace, so the signal had settled before any other device on the PCB needed to use the voltage as an input. The propagation time for an electrical signal on a PCB is on the order of two nanoseconds for every foot of trace. Now that very fast devices can turn on or off very quickly, these transmission line effects can ruin a design if they are not understood.

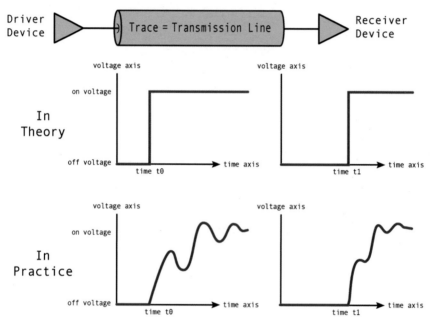

Figure 46. Transmission line effects on a PCB trace

An example of a potentially dangerous situation is shown in Figure 46. A digital device turns on, driving a trace to a high voltage. The top timing diagrams show what the designer desires. The device turns on instantaneously. A short time later, the device at the end of the trace sees this clean voltage change. What is more likely to happen is shown in the timing diagrams at the bottom of the figure. The driver begins to turn on. As it does so, a wave propagates down the trace and reflects back, creating bumps in the output of the device so that it is no longer a nice clean signal. At the receiver, these waves create even larger bumps in the voltage levels. If these bumps are extreme enough, the receiver device will behave as though the driver is turning on and off several times. This can cause a digital circuit to behave incorrectly.

There are several possible ways to handle the transmission line effect for critical traces. The particular solution will depend on the particular design. Some guidelines are:

a) Critical traces should not have any stubs. In other words, these traces should be, as much as possible, straight lines. Traces that branch off in different directions create multiple waves that interfere with each other in unpredictable ways.

b) Critical traces should be as short as possible.

c) The receivers on critical traces should be grouped together so that they all see the same signal. If they see different inputs, their outputs will not correspond to each other, resulting in unpredictable outputs.

d) Serial termination. This involves putting resistors between the drivers and the trace. This slows down the speed at which the drivers change voltage. A slower change in voltage will not tend to produce waves.

e) Parallel or Thevenin termination. This involves putting resistors at the end of the trace, connected to power and ground. This dampens any reflecting wave, reducing its voltage.

Transmission Lines

Quiz 5: Transmission Lines

1. **Transmission lines are any conductor that acts like a combination of many resistors, capacitors, and inductors.**

   ☐ True ☐ False

2. **All conductors are transmission lines.**

   ☐ True ☐ False

3. **Only high speed conductors act as transmission lines.**

   ☐ True ☐ False

4. **Reflection and absorption of electrical signals at the end of the transmission line causes electrical waves in the line.**

   ☐ True ☐ False

5. **A wave propagates on a printed circuit board (PCB) trace at the rate of:**

   ☐ ten miles per hour ☐ the speed of light

   ☐ the speed of sound ☐ about 1 foot per 2 nsec

6. **Transmission line effects on a PCB trace cause smoother, more reliable signals.**

   ☐ True ☐ False

7. **Critical PCB traces should be, as much as possible, straight lines.**

   ☐ True ☐ False

8. **Critical PCB traces should be as long as possible.**

   ☐ True ☐ False

9. **The receivers on critical PCB traces should be spread apart.**

   ☐ True ☐ False

10. **Serial termination involves**
    - ☐ putting resistors between the drivers and the trace
    - ☐ putting capacitors between the drivers and the trace
    - ☐ putting resistors between the receivers and the trace
    - ☐ putting transistors between the drivers and the trace
11. **Parallel or Thevenin termination involves**
    - ☐ putting resistors at the beginning of the trace, connected to power
    - ☐ putting resistors at the beginning of the trace, connected to ground
    - ☐ putting resistors at the end of the trace, connected to power and ground
    - ☐ putting resistors at the end of the trace, connected to another trace
12. **Power lines to your house are transmission lines.**
    - ☐ True   ☐ False

7. Digital Logic

7.1 Boolean Algebra

Boolean algebra was developed in 1849 by British mathematician George Boole. It was intended as method of representing logical thought and reason in terms of equations. George Boole probably never realized that his new algebra would become the basis of an industrial revolution many years later. It was the brilliant American mathematician Claude Shannon who rediscovered Boole's work nearly 100 years later and recognized the utility of Boolean algebra for designing computer circuits.

Boolean algebra deals with statements that are either true or false, based on whether the facts they are about are true or false. Suppose we take the following statements.

    A.   Bob is the author of this book. (TRUE)

    B.   This book is about engineering. (TRUE)

Now let us evaluate whether the following statement is true or false.

    C.   Bob is the author of this book and it is not about engineering.

Boolean logic says that we can rewrite the previous statement as

    C = A AND NOT B

A and B are called the "operators" while "AND" is the operation. If we substitute TRUE and FALSE where they belong, we come up with this.

    C = TRUE AND NOT TRUE

or

Digital Logic

   C = TRUE AND FALSE

For a statement with an AND to be TRUE, both parts need to be true. Therefore, we know that statement C is false.

   C = FALSE

Mathematicians, scientists, and engineers prefer to use symbols and numbers rather than words, because they take up less space (not because they confuse outsiders). They use the number 0 to represent something that's FALSE and the number 1 to represent something that's TRUE. The symbol & is used for AND, the symbol | is used for OR, and the symbol ~ is used for NOT[3]. Rewriting the statements above using symbols, we get

   A = 1

   B = 1

   C = A & ~B = 0

Suppose we look at a statement that uses an OR. Now let us evaluate whether this statement is true or false.

   A.   Bob is the author of this book. (TRUE)
   B.   John is the author of this book. (FALSE)
   C.   Bob is the author of this book or John is the author of this book.

Boolean logic says that we can rewrite the previous statement as

   C= A OR B

---

[3] These are not the only symbols that are used to represent these values, operators, and operations. Often electrical engineers will use different symbols than mathematicians or scientists to represent the same exact things.

Again A and B are called the "operators" and "OR" is the operation. If we substitute TRUE and FALSE where they belong, we come up with this.

    C = TRUE OR FALSE

For a statement with an OR to be TRUE, only one part needs to be true. Therefore, we know that in this case, statement C is true.

    C = TRUE

These are the basic operations of Boolean algebra: AND, OR, and NOT. The tables for the AND and OR operations are shown in Figure 47 below. The table on the left side of the figure shows how to understand the other tables. The operation is shown in the top left corner. Find the value of one operator on the left side (0 or 1). That gives you the row to look at. Then find the value of the other operator on the top (0 or 1). That gives you the column. Where the row and column meet, shows you the value of the entire statement. For example, using the AND table, 0 on the left and 1 on the top meet at a 0 value. This means that 0 & 1 (FALSE AND TRUE) results in 0 (FALSE). Using the OR table, 0 on the left and 1 on the top meet at a 1 value. This means that 0 | 1 (FALSE OR TRUE) results in 1 (TRUE).

All digital logic design, and nearly all computers in existence, are based on these three simple Boolean operators AND, OR, and NOT. However, more operators have been added since Boole created them. Sometimes new operators make equations simpler. Sometimes new operators correspond more closely to how the electronic circuitry operates that implement these operations. In any case, all digital logic can be designed with only the basic three operators and a timed delay, which allows sequences of logic to take place.

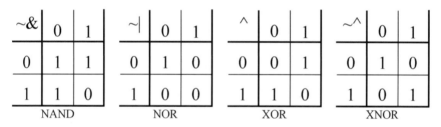

Figure 47. Basic Boolean operators

Some common additional operators are not OR (called a NOR operator), not AND (NAND), exclusive OR (XOR), and exclusive NOR (XNOR). These additional operators are shown in Figure 48.

Figure 48. Additional Boolean operators

7.2 Binary Numbers

The next important concept in digital logic is the binary number system. The number system that we use for everyday calculations—counting, making change, weighing ourselves, etc.—are base ten, or decimal, numbers. This means that they are based on the number ten. As we know from elementary school, every number has a ones column, a tens column, a hundreds column, etc. as needed. Every time we count to ten, we add one to the tens column of a number and start counting over again. When we have 10 tens in the tens column, we add one to the hundreds column, etc.

The same is true for binary numbers, except that they are base two instead of base ten. This means we when count to 2, we add one to the twos column and start over. When we get 2 twos, we add one to the fours column and start over again. Counting in decimal and binary is shown in Table 1. As you can see, binary numbers have many more bits ("binary digits") than the equivalent decimal numbers. So why do we use them to design digital logic? This is because it is much easier to determine whether a circuit is on or off (1 or 0) than whether the circuit is producing one of ten voltage levels. The transistor, discussed earlier, is an ideal switch for implementing the 0 or 1 required for binary numbers.

Decimal Counting	*Binary Counting*
0	0000
1	0001
2	0010
3	0011
4	0100
5	0101
6	0110
7	0111
8	1000

Table 1. Counting in decimal and binary

With the use of binary numbers, digital logic is able to do very advanced calculations in addition to logical operations described above.

7.3 Schematic Representations

When Boolean logic functions are implemented in an electrical circuit, transistors are used, along with resistors, to create the appropriate function. The operands are input signals, while the results are output signals. These input and output signals have either a low voltage (0 or false) or a high voltage (1 or true). Rather than require engineers to draw all of the transistors and resistors on a schematic, a simplified method was developed. The symbols that represent each common logic function are shown in Figure 49. Simple circuits that implement basic

Digital Logic

Boolean logic are referred to as gates.

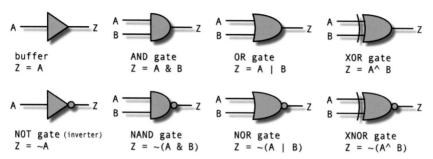

Figure 49. Boolean logic gates

7.4 Delay Elements

The next important element to digital design is the delay element. Boolean logic allows circuits to implement any logic function that is a combination of the inputs. By representing numbers in binary, Boolean logic allows calculations to take place. Delay elements are the final ingredient. They allow digital logic to implement algorithms that need to take place in a particular order. Complex step-by-step procedures can be implemented in order to perform calculations or control complex machinery.

The most common delay element is a flip-flop, shown in Figure 50. The clock signal is an input that rises to a high voltage, then drops to a low voltage at regularly repeating intervals. The clock is used to synchronize every flip-flop in the system and maintain constant intervals. The input signal goes to the D input pin of the flip-flop. When the input signal changes from a 0 to a 1 or from a 1 to a 0, the outputs are unchanged until after the clock goes from a 0 to 1. Only after this "rising edge" of the clock do the outputs change. The Q output changes to the value of the D input. The QN output changes to the opposite value of the D input. In the diagram, the D input is connected to a signal called IN, the CLK input is connected to a signal called CLOCK, and the Q and QN outputs are connected to signals labeled OUT and OUTN respectively.

Just Enough Electronics to Impress Your Friends and Colleagues

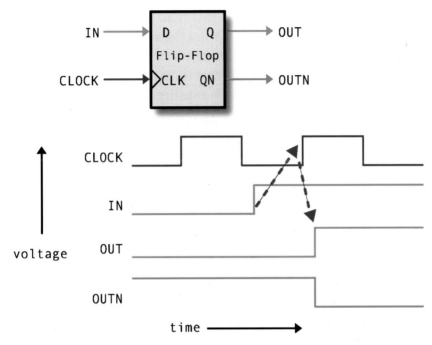

Figure 50. Flip-flop

7.5 Finite State Machines

Finite state machines, or simply state machines, are combinations of Boolean logic, also called combinatorial logic, and clocked elements, typically flip-flops. By putting these two types of logic together, any form of algorithm can be implemented. All digital logic consists, essentially, of state machines used to implement a wide variety of functions.

A state diagram is a way of describing the actions of a state machine using simple figures. A state diagram for a simple state machine is shown in Figure 51. It may be the state machine for a chip that can read or write memory. The machine starts out in the IDLE state. It remains in the IDLE state until the GO signal is asserted (goes to a true level). Then, if the WR signal is asserted, the machine enters the WRITE state, after which it returns to the IDLE state and the action begins again. If the WR signal is not asserted, however, the machine enters the READ1 state. If the controller is doing a fast read and the DONE signal is asserted by some other logic, then the state machine returns to the IDLE state. Otherwise it goes into the READ2 state and remains there until the DONE signal is asserted. The machine then returns to the IDLE state. The status of each signal is checked only on the rising edge of a clock signal that is used to time the circuit and synchronize it to every other circuit on the chip or in the system. The speed of the state machine is controlled by the frequency of the clock signal.

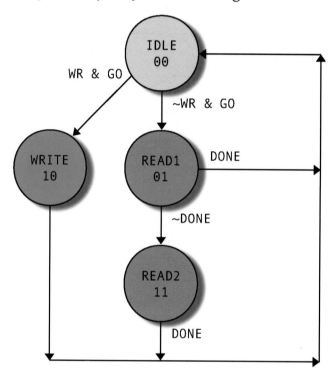

Figure 51. Simple state machine

The numbers under the state names in each circle represent the encoding of the states when the state machine is implemented using logic and flip-flops. There are many ways of encoding states, but this one was the one chosen for this particular implementation of this state machine. The implementation is shown in Figure 52. Note that the two bits representing each state in the diagram correspond to signals S1 and S0.

State machines are the essence of digital design. As mentioned before, all algorithms, control functions, and calculation circuits can be implemented as state machines.

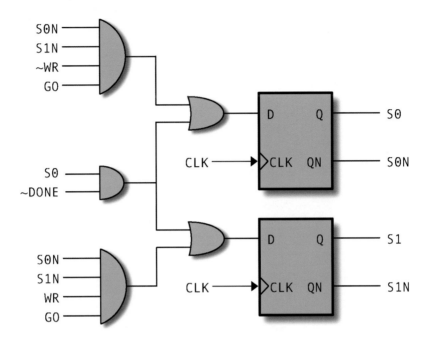

Figure 52. State machine implementation using logic gates and flip-flops

Digital Logic

Quiz 6: Digital Logic

1. **Given the following three statements and their logic values, determine whether the next statements are true or false according to Boolean logic.**

    A. The sun is shining (TRUE)

    B. This earth is flat (TRUE)

    C. Everybody cares about science (FALSE)

    ☐ TRUE   ☐ FALSE   The earth is flat and the sun is not shining.

    ☐ TRUE   ☐ FALSE   The earth is not flat or the sun is not shining.

    ☐ TRUE   ☐ FALSE   The earth is flat and not everybody cares about science.

    ☐ TRUE   ☐ FALSE   Everybody cares about science or the sun is shining.

2. **Given the following three logic values, determine whether the next Boolean logic statements are true or false.**

    A.   TRUE

    B.   TRUE

    C.   FALSE

    ☐ TRUE   ☐ FALSE   B AND NOT A

    ☐ TRUE   ☐ FALSE   NOT B OR NOT A

    ☐ TRUE   ☐ FALSE   B AND NOT C

    ☐ TRUE   ☐ FALSE   C OR A

Digital Logic

3. Given the following three Boolean values, determine whether the next Boolean equations equal 1 or 0.

    A. 1

    B. 1

    C. 0

    B & ~A =

    ~B | ~A =

    B & ~C =

    C | A =

4. Which of the following are basic Boolean operators?

    ☐ NAND ☐ OR ☐ PLUS ☐ NOR

    ☐ BUT ☐ AND ☐ NOT ☐ XOR

    ☐ XNOR ☐ MAYBE ☐ ALSO ☐ IF

5. What is the number base for decimal numbers?

    ☐ 1 ☐ 2 ☐ 10 ☐ 12

6. What is the number base for binary numbers?

    ☐ 1 ☐ 2 ☐ 10 ☐ 12

7. Delay elements are implemented in digital logic using

    ☐ flip-flops ☐ floppers ☐ thongs ☐ delay lines

Just Enough Electronics to Impress Your Friends and Colleagues

8. Match the schematic symbol with its Boolean function.

(NOR gate symbol)	AND
(NAND gate symbol)	OR
(NOT gate symbol)	NOT
(AND gate symbol)	NAND
(OR gate symbol)	NOR
(NOT/inverter symbol)	XOR
(XNOR gate symbol)	XNOR

Digital Logic

9. **Fill in the missing numbers from the binary counting sequence.**

   000
   001
   010
   ___
   100
   101
   ___
   ___

10. **Calculate the answer to the following addition problem in binary**

      010
   + 011

11. **Calculate the answer to the following multiplication problem in binary**

      010
   x 011

12. **Finite state machine**
   - ☐ are combinations of Boolean logic and clocked elements
   - ☐ are typically implemented with Boolean logic gates flip-flops
   - ☐ can implement any form of algorithm
   - ☐ are often designed using state diagrams
   - ☐ all of the above

8. Semiconductor Technology

In this section we discuss the characteristics of a semiconductor, how semiconductors are used to make diodes and transistors, and how semiconductors are combined to make integrated circuits.

8.1 Conductors, Insulators, and Semiconductors

A conductor is a material that, when a voltage is applied, allows current to pass through it. A conductor has a low resistance value. An insulator has a high resistance value, such that very little current, if any, passes through it when a voltage is applied. A semiconductor, an amazing discovery of the twentieth century, acts as a conductor in some situations and an insulator in others.

An atom, as we learned in school, is composed of a small core called a nucleus consisting of uncharged neutrons and positively charged protons. Zipping around the nucleus are negatively charge electrons. The total charge of an atom is zero. There are exactly equal numbers of protons and electrons.

Some atoms have electrons that are not very strongly coupled to the nucleus. These atoms are known as conductors, since a voltage can be applied to them causing electrons to get knocked off the atom. These electrons hop from one atom to the next, creating an electric current.

Some atoms have electrons that are very strongly attracted to the nucleus. These atoms are considered insulators, because the electrons do not easily leave the atom. Only an extremely high voltage will knock electrons off the atom.

Some atoms, on the other hand, have a group of electrons that are very strongly held in place by the nucleus and another group of electrons that are easy to knock off. These are the semiconductors; specifically, these are intrinsic semiconductors. Extrinsic semiconductors are obtained by combining two different elements to create a material that has the properties of a semiconductor. Silicon is an example of an intrinsic semiconductor. Gallium arsenide, which is a combination of gallium and arsenic, is an extrinsic semiconductor.

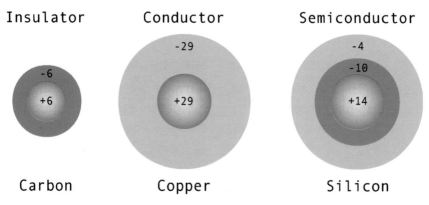

Figure 53. Insulators, conductors, and semiconductors

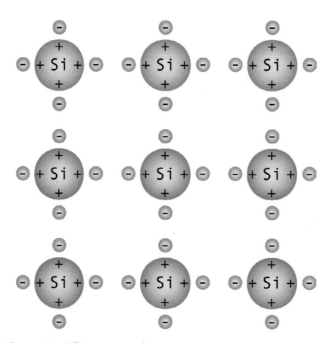

Figure 54. Silicon crystal

Semiconductors tend to form very regular crystals with regular structures and nice, well-defined pathways for the electrons to follow. The "loose" electrons tend to wander throughout the crystal. Each silicon atom, for example, has four loose electrons, as shown in Figure 54.

In a process called doping, impurities can be added to the semiconductor crystal lattice in a controlled way. If the impurity atom has more loose electrons than the base semiconductor, as shown in Figure 55, then the impurity is called a donor and the resulting semiconductor is called an n-type semiconductor because it has extra negative electrons floating around. If the impurity atom has fewer loose electrons than the base semiconductor, as shown in Figure 56, then the impurity is called an acceptor and the resulting semiconductor is called a p-type semiconductor because it has fewer negative electrons floating around.

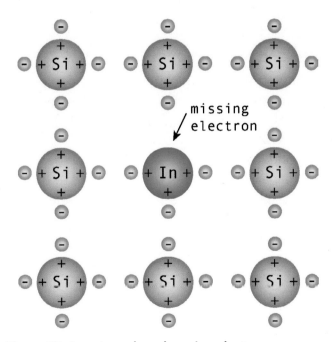

Figure 55. An n-type doped semiconductor

Semiconductor Technology

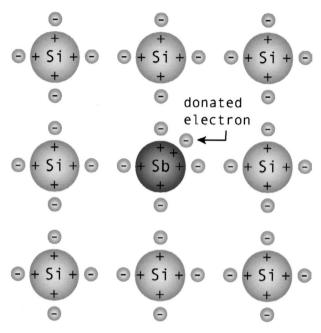

Figure 56. A p-type doped semiconductor

8.2 Diodes and pn Junctions

The interesting effects of semiconductors happen when p-type semiconductors are placed next to n-type semiconductors, creating what is called a pn junction, shown in Figure 57. When the cathode (positive terminal) of a battery is connected to the p-type semiconductor region of the device and the anode (negative terminal) of the battery is connected to the n-type semiconductor region, current will flow. When the battery is reversed, no current flows. This creates the simple, but important nonlinear device, the diode that we discussed earlier.

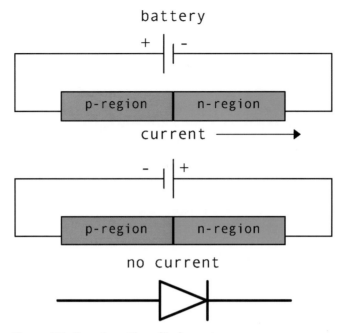

Figure 57. A pn junction diode

8.3 Transistors and npn and pnp Junctions

Now that we have seen some interesting stuff by connecting differently doped semiconductor materials next to each other, let's extend it. What happens when we place three materials together in an npn configuration as shown in Figure 58? Normally, the top n-region will resist sending electrons to the positive terminal of the battery (the cathode), although the n-side at the bottom will prefer to receive electrons from the negative terminal of the battery (the anode). The effect is that current does not flow. However, applying a very small positive voltage to the p-region creates a very small current through the bottom pn junction diode. This small "base current" actually coaxes the current to move through the circuit on the right through the entire npn structure. This is exactly the transistor that we discussed earlier. The circuit on the left can control the circuit on the right.

Semiconductor Technology

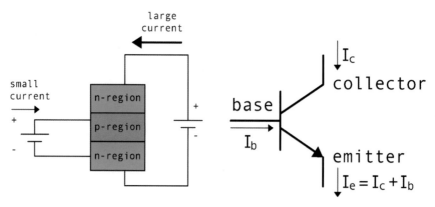

Figure 58. An npn transistor

8.4 Integrated Circuits

Once the technology was invented to create a transistor, the next goal was to place as many as possible of these devices together to create more complex devices. The result was the integrated circuit, a chip of silicon with impurities implanted in specific areas and different materials placed next to each other and on top of each other to create entire circuits. The credit for this important invention is shared by Jack Kilby, working at Texas Instruments, and Robert Noyce, a founder of Intel Corporation. Their creations were done independently and nearly simultaneously. Jack Kilby was awarded the 2000 Nobel Prize for Physics for this invention. Unfortunately Robert Noyce had passed away by that time, and the Nobel Prize is only awarded to living scientists. A cross-section illustration of such an integrated circuit (IC) is shown in Figure 59, while an actual photograph of an IC is shown in Figure 60 where each line has a width of 0.5 microns, roughly 2000 times smaller than an average grain of sand.

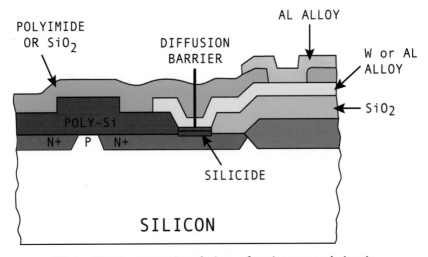

Figure 59. Cross sectional view of an integrated circuit

Figure 60. Scanning electron microscope picture of an IC

These integrated circuits typically start out as a thin slice of p-type silicon called a wafer. There are then a series of process steps that are performed. These process steps can be one of two types. One type of process places a material into the wafer as a dopant or onto the wafer as a new layer. The second type of process removes undesired materials from the wafer. Methods for applying materials include ion implantation, oxidation, and chemical vapor deposition. Processes used to remove material from a wafer include vapor phase etching, chemical etching, and dissolution. These processes are repeated many times, depositing and removing layers of materials in particular patterns. Finally, a thin protective coating of a material, such as silicon dioxide, is applied to the entire wafer in order to seal it and protect it from damage. The wafers are then diced into individual chips and packaged into the rectangular packages with which we are familiar.

8.5 Growing Silicon

Figure 61. Silicon boules

To obtain a wafer in order to begin the fabrication process, a silicon crystal is grown from a seed that is drawn slowly from a vat of liquid silicon. This process is called the Czochralski method, named after Polish scientist Jan Czochralski who invented the method in 1916 while investigating the crystallization rates of metals. The liquid silicon is maintained at a temperature slightly above its melting point so that the silicon solidifies as it is drawn out and cooled. The silicon crystal is called a boule and has a gray, shiny texture as can be seen in Figure 61. The crystal is cut into thin slices, each about 300 microns (μm) thick, like the one shown in Figure 62, that are then polished to ensure that the faces of the wafers are parallel and free from contamination.

8.6 Oxidation

Silicon dioxide is used as an insulator to keep circuits electrically isolated. The oxidation procedure involves exposing silicon to oxygen at about 1000°C. Just like with the formation of rust (iron oxide) on iron, the layer of silicon exposed to oxygen chemically combines with it to form silicon dioxide.

Figure 62. A blank silicon wafer

8.7 Lithography and Photoresist

Photolithography is a process in which a wafer is coated with a light sensitive material called photoresist. The photoresist is first deposited on the wafer. This can be done by spraying the wafer with the liquid resist (chemical deposition) or blowing the vapor photoresist over the wafer (gas deposition). A mask, like a stencil, is placed on top of the wafer and it is exposed to ultraviolet (UV) light. If the photoresist is a positive photoresist, then where the light hits the resist, it fixes it to the wafer. If it is a negative photoresist, then the light breaks it down so that it does not affix to the wafer. The wafer is then washed with a developer solution to remove the photoresist that is not bound to the wafer.

8.8 Ion implantation and Diffusion

Ion implantation involves accelerating a beam of ions using a high voltage electric field, similar to the way a television cathode ray tube (CRT) accelerates a beam of electrons to draw a picture on the screen. The ions penetrate through the oxide layer and enter the silicon, becoming dopants, changing the electrical characteristics of the silicon. The ion beam will not penetrate areas covered by the photoresist. This allows areas of implantation to be tightly controlled. The barrage of the ion beam causes the regular silicon crystal to be damaged. An annealing process is used to repair the crystal and diffuse the ions into it. Annealing involves heating the crystal, then slowly bringing it back to room temperature. This allows the crystal atoms and ions to jiggle around and settle back into place, like shaking a cereal box to mix in the raisins and get it all evenly distributed in the box.

8.9 Etching

Etching is the process whereby material is selectively removed from the wafer. Chemical etching involves dissolving the material using a chemical solution. Physical etching involves "sandblasting" an area with high-energy ions.

Finally, when all of the processes have completed in the proper order, the resulting wafer, covered with die that will be eventually cut into chips for packaging, may look something like the picture in Figure 63.

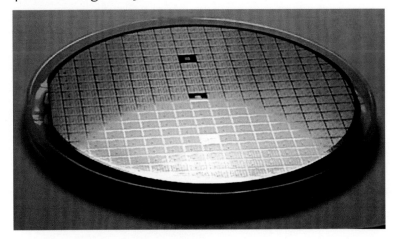

Figure 63. A finished wafer

Semiconductor Technology

Quiz 7: Semiconductors

1. Match the term to its correct definition.

a) Conductor	Does not allow a significant amount of electricity to flow through it when a voltage is applied.
b) Insulator	Allows a large amount of electricity to flow through it when a voltage is applied.
c) Semiconductor	Under some conditions, does not allow much electricity to flow though it. Under other conditions, allows a large amount of electricity to flow through it.

2. Check all three basic parts of an atom.
   - ☐ electron
   - ☐ tachyon
   - ☐ proton
   - ☐ neutron
   - ☐ positron
   - ☐ moron

3. A material with a very regular atomic structure, where the atoms are all lined up, is called
   - ☐ an atom
   - ☐ a solid
   - ☐ a molecule
   - ☐ a crystal
   - ☐ a liquid
   - ☐ a semiconductor

4. The most common semiconductor material used to create integrated circuits is
   - ☐ silicon
   - ☐ carbon
   - ☐ copper
   - ☐ plastic
   - ☐ germanium
   - ☐ hydrogen

5. When we "dope" a silicon crystal with atoms that have more free electrons than the silicon does, the resulting substance is called
   - ☐ n-type semiconductor
   - ☐ insulator
   - ☐ conductor
   - ☐ p-type semiconductor

Semiconductor Technology

6. When we "dope" a silicon crystal with atoms that have fewer free electrons than the silicon does, the resulting substance is called

    ☐ n-type semiconductor   ☐ insulator

    ☐ conductor   ☐ p-type semiconductor

7. A junction of p-type semiconductor material and n-type semiconductor material is called a pn junction, or

    ☐ a transistor   ☐ an integrated circuit

    ☐ a resistor   ☐ a diode

8. A sandwich of n-type, p-type, and n-type semiconductor materials is called

    ☐ a transistor   ☐ an integrated circuit

    ☐ a resistor   ☐ a diode

9. A semiconductor device that is made up of many transistors is called a

    ☐ transistor   ☐ integrated circuit

    ☐ resistor   ☐ diode

10. A tube of silicon drawn out from a vat of liquid silicon and cooled is called a

    ☐ boule   ☐ die

    ☐ wafer   ☐ semiconductor

11. A round, polished slice of silicon is called a

    ☐ boule   ☐ die

    ☐ wafer   ☐ semiconductor

12. **Which one of the following is not a step in the process of creating silicon wafers?**

    ☐ oxidation   ☐ lithography

    ☐ residualization   ☐ etching

    ☐ ion implantation   ☐ diffusion

    ☐ photoresist   ☐ Czochralski method

Semiconductor Technology

9. Memory Devices

Memory devices store bits of information. They generally fall into one of four categories – Read Only Memories (ROMs), Nonvolatile Programmable Memories (PROMs, EPROMs, EEPROMs, NVRAMs), Static Random Access Memories (SRAMs), and Dynamic Random Access Memories (DRAMs). Each of these types of memories is discussed below. As always, newer types of memories are becoming available. Which ones will become standards in the future remains to be seen. In any case, these four basic types will still be around for a long time to come.

9.1 Read Only Memories (ROMs)

As the name implies, ROMs contain bit patterns that are hardwired into the device during production and cannot be changed. An advantage of ROMs is that they do not lose data under any situation. All other devices, including Programmable ROMs that can be programmed by the user to contain specific data, will eventually wear out and lose data, though after many uses or many years. ROMs are also very inexpensive to produce in large quantities since they have a simple fabrication process and require no additional programming after production. A disadvantage is that they cannot be changed. If new data is required, the ROM must be physically removed from the circuit board and replaced with a new ROM. Also, in small quantities, ROMs are expensive since they are custom produced for one customer only.

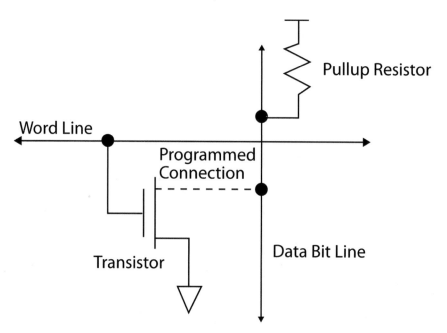

Figure 64. ROM single bit cell

The basic diagram for a ROM cell containing a single bit of data is shown in Figure 64. The word line is turned on if the address into the chip includes this particular bit cell. The metal layer is used to program the data into the ROM during fabrication. In other words, if the metal layer mask has a connection between the transistor output and the data line (shown as the "programmed connection" in the figure), the bit is programmed as a zero. When the bit is addressed, the output will be pulled to a low voltage, a logic 0. If there is no connection, the data line will be pulled up by the resistor to a high voltage, a logic 1.

9.2 Nonvolatile Programmable Memories

Nonvolatile programmable memories are memory devices that can be programmed with specific data after they have been manufactured. They can usually be erased and reprogrammed with new data also. The nonvolatile term means that they do not lose their data in normal operation, even after power is turned off to them.

9.2.1 Programmable Read Only Memories (PROMs)

The basic concept of the Programmable Read Only Memory, or PROM, was invented in 1956 by American engineers Wen Tsing Chow and William Henrich, working for the Arma Division of the American Bosch Arma Corporation, for which they were granted U.S. patent 3,028,659 in 1962. That PROM was a circuit on a board that consisted of a matrix of wires and diodes. Modern PROMs are programmable one time only and consist of an array of fuses or antifuses on a semiconductor chip. Fuses, like household fuses, consist of a wire that breaks connection when enough current goes through it. Antifuses normally have gaps in the metal connection. If enough voltage is applied, metal actually migrates between the two terminals to create a connection. A diode based PROM cell is shown in Figure 65, while a transistor based PROM cell is shown in Figure 66.

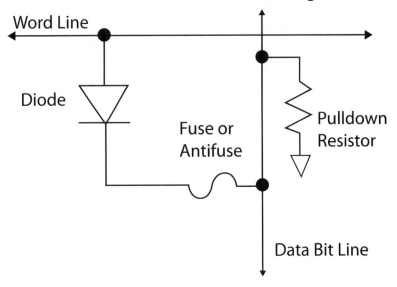

Figure 65. Diode based PROM cell

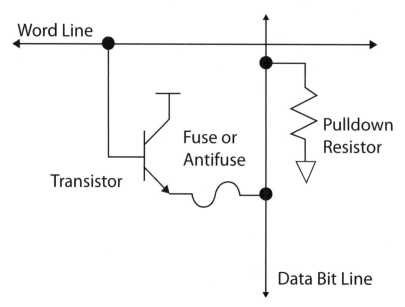

Figure 66. Transistor based PROM cell

9.2.2 Reprogrammable Read Only Memories (EPROMs, EEPROMs, Flash, and NVRAMs)

Reprogrammable memories like EPROMs, EEPROMs, flash memories, and NVRAMs, essentially trap electric charge on the input of a transistor that is not connected to anything. The input acts like a capacitor. The transistor amplifies the charge, while a second transistor turns on when the bit is being addressed. During programming, the charge is injected onto the transistor by one of several methods including tunneling and avalanche injection. This charge will eventually leak off. In other words, some electrons will gradually escape, but the leakage will not be noticeable for a long time, on the order of ten years, so that they remain programmed even after power has been turned off to the device. Programming one of these devices causes wear and tear on the chip while the electrons are being injected. Most of these devices can be programmed about 100,000 times before they begin to lose their ability to be programmed.

9.2.2.1 Erasable Programmable Read Only Memory (EPROM)

Invented by Dov Frohman-Bentchkowsky of Intel Corporation in 1971, for which he was awarded U.S. patent 3,660,819 in 1972, EPROMs must be erased before they can be reprogrammed with new data. They can be erased in a number of different ways, depending on the device. An Erasable Programmable Read Only Memory (EPROM) is programmed electrically, in an electric circuit using a high voltage, but is erased by exposing it to ultraviolet light for a period of several minutes. The UV light causes the electrons on the transistors to dissipate. The obvious disadvantage to this is that in order to reprogram a device it must be removed from the circuit and placed in a UV light source. A picture of a packaged EPROM memory can be seen in Figure 67 showing the window that allows UV light to get through.

Figure 67. EPROM package

9.2.2.2 Electrically Erasable Programmable Read Only Memory (EEPROM)

Phillip Salsbury and George Perlegos of Intel Corporation

invented the Electrically Erasable PROMs (EEPROMs) in 1975 and were granted U.S. Patent 3,938,108 in 1976. EEPROMs are programmed electrically and erased electrically. These can be reprogrammed without removing them from the circuit board. These devices use two transistors per single bit cell.

9.2.2.3 Flash Memory

EPROMs and EEPROMs tend to have very slow access times. Flash EEPROMs, or simply flash memories, use only a single transistor per cell, making them much faster than traditional EPROMs and EEPROMs. These devices were invented by Japanese engineer Fujio Masuoka of Toshiba Corporation in 1984. Unlike other kinds of programmable memories, EEPROM Flash memories are used in high speed circuits. They are called "flash" because bulk portions of the entire device can be erased very quickly in one "flash." Figure 68 shows one of the many USB flash devices that have become ubiquitous for computer users in recent years to transfer files and backup data.

Figure 68. Packaged flash memory

9.2.2.4 Non Volatile Random Access Memory (NVRAM)

A Non Volatile Random Access Memory (NVRAM) is actually an SRAM or DRAM as described in the following sections. The difference is that the NVRAM has a way to store its data even when power is turned off. Some NVRAMs have an EEPROM bit for every RAM bit. When power is removed from the device, the EEPROM bit saves the RAM bit value. When the device is powered on again, the entire contents of the device are loaded back into the RAM cells from the EEPROM cells. In this way the device has the characteristics of a RAM with the programmability and nonvolatility of a PROM. Other NVRAMs have a small battery connected to them so that when the system's power is turned off the NVRAM is still powered on. For those of us old enough to remember the older personal computers, the BIOS that allowed the computer to boot up used a small battery to keep its contents alive while the computer was powered down.

9.3 Dynamic Random Access Memories (DRAMs)

Invented in 1968 by Robert Dennard at IBM Corporation, DRAMs are the workhorses of memory devices. Dennard was granted U.S. Patent 3,387,286. DRAMs use a single transistor per bit cell to charge or discharge a capacitor in order to create a high or low voltage that translates into a logic 1 or 0. A schematic of this cell is shown in Figure 69. A DRAM can be written and read any number of times, indefinitely, while the system is powered on. When power is removed, the DRAM loses all data.

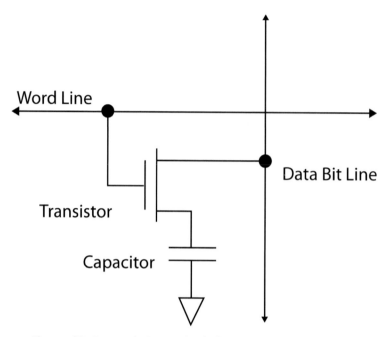

Figure 69. Dynamic RAM single bit cell

An advantage of DRAMs is that they can have a very high density because each bit requires only one very small transistor. They are also relatively inexpensive, mainly due to the fact that they are produced in such large quantities. In addition they do not consume much power. A disadvantage is that they are slower than SRAMs, described in the next section, and also, the capacitive charge must be refreshed periodically because otherwise it will leak off in a very short time on the order of seconds. The DRAM has a refresh circuit on the chip. The engineer must design a circuit that asserts the input signals in a particular sequence, which informs the on-chip refresh circuit that it is time to refresh the chip. The refresh circuit then internally boosts up the charge on all charged capacitors. The system cannot access the DRAM during this refresh period. A diagram of an array of DRAM bits inside a DRAM chip is shown in Figure 70.

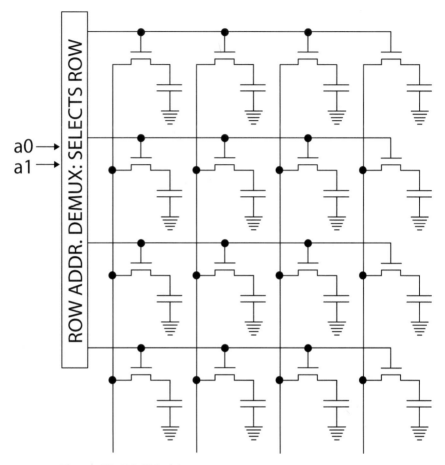

Figure 70. DRAM chip array

Those of us who have upgraded the memory in our personal computers will recognize the DRAM module in Figure 71 that holds several DRAM chips and plugs into the computer's memory bus.

Figure 71. DRAM module

9.4 Static Random Access Memories (SRAMs)

If DRAMs are the workhorses, SRAMs are the racecars of memory devices. SRAMs were introduced by Intel Corporation in 1969. Like DRAMs they can be read and written any number of times while power is applied. Once power is turned off, the SRAM forgets its data.

SRAMs use two transistors or more per bit cell. The extra transistors are used to latch the stored electric charge so that it does not leak away. This is kind of like a two-person juggling act with each transistor supplying electrons to the other. This gives the first advantage of SRAMs over DRAMs – they do not need to be refreshed because they are constantly refreshing themselves. Another advantage is that they are much faster than DRAMs. The disadvantages are that they are much more expensive and consume much more power. Also, due to the number of transistors required for each bit, they are not as dense as DRAMs. In other words, they cannot hold as many bits in a device of the same physical size. One kind of DRAM cell is shown in Figure 72.

Just Enough Electronics to Impress Your Friends and Colleagues

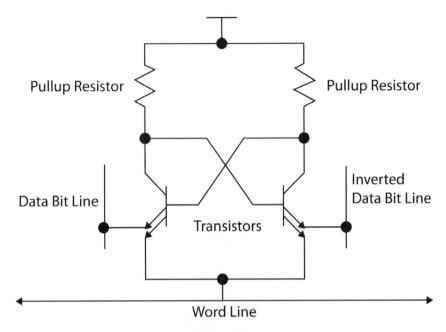

Figure 72. Static RAM single bit cell

9.5 Memory Organization

When you select a memory device, whether it is a DRAM, SRAM, PROM, etc., the package looks something like that in Figure 73. There are two signals, WR and RD, which control whether the data is being written to the device or read from it. For most devices, it is not allowed to write and read at the same time, so the behavior of the memory is unpredictable if both signals are asserted (at a logic 1 voltage) at the same time. Some devices cannot be written, so there is no WR input signal. Some more complex devices have additional signals, but all of them are essentially similar in functionality to this one shown.

There are also address bits that specify which location in the memory we are writing or reading. For the device shown, there are 6 address bit inputs. All of them can be either a 0 or 1. That means that there are 2^6 possibilities, or 64 locations in this memory. Each location holds 8 bits of data, since there are 8 data signals. So we say that this is a 64 by 8 memory device. In order to group bits of data together, someone came up with the description of a byte, which is simply 8 bits of data that are grouped together. So another way of describing this memory is that it holds 64 bytes of data.

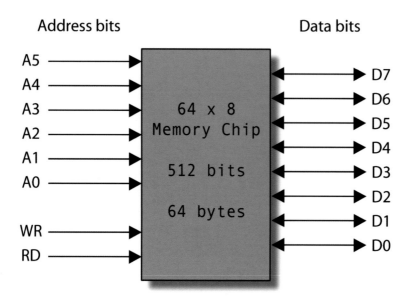

Figure 73. Memory chip diagram

Quiz 8: Memories

1. Which of the following memories lose data when they are powered down? (check all that apply)

    ☐ ROM   ☐ PROM   ☐ EPROM   ☐ EEPROM

    ☐ NVRAM   ☐ SRAM   ☐ DRAM

2. Which of the following memories can have its data overwritten? (check all that apply)

    ☐ ROM   ☐ PROM   ☐ EPROM   ☐ EEPROM

    ☐ NVRAM   ☐ SRAM   ☐ DRAM

3. Which of the following memories cannot be erased? (check all that apply)

    ☐ ROM   ☐ PROM   ☐ EPROM   ☐ EEPROM

    ☐ NVRAM   ☐ SRAM   ☐ DRAM

4. A ROM bit is programmed with (check all possibilities)

    ☐ a transistor and a fuse

    ☐ a metal connection and a transistor

    ☐ electric charge on the input of a transistor

    ☐ a diode and an antifuse

5. A PROM bit is programmed with (check all possibilities)

    ☐ a transistor and a fuse

    ☐ a metal connection and a transistor

    ☐ electric charge on the input of a transistor

    ☐ a diode and an antifuse

Memory Devices

6. An EPROM bit is programmed with (check all possibilities)
   - ☐ a transistor and a fuse
   - ☐ a metal connection and a transistor
   - ☐ electric charge on the input of a transistor
   - ☐ a diode and an antifuse

7. An NVRAM can be a combination of (check all possibilities)
   - ☐ DRAM and EEPROM
   - ☐ PROM and battery
   - ☐ DRAM and SRAM
   - ☐ SRAM and battery
   - ☐ PROM and SRAM
   - ☐ SRAM and EEPROM

8. Which of the following memories generally has the highest density of bits?
   - ☐ SRAM
   - ☐ DRAM

9. Which of the following memories needs to be regularly refreshed?
   - ☐ SRAM
   - ☐ DRAM

10. Which of the following memories is generally fastest?
    - ☐ SRAM
    - ☐ DRAM

11. A 16 by 4 memory has _____ address lines and _____ data lines.

12. A 128 byte memory has _____ address lines and _____ data lines.

10. Application Specific Integrated Circuits (ASICs)

An Application Specific Integrated Circuit, or ASIC, is a chip that can be designed by an engineer with no particular knowledge of semiconductor physics or semiconductor processes. The ASIC vendor has created a library of cells and functions that the designer can use without needing to know precisely how these functions are implemented in silicon. The ASIC vendor also provides or supports third-party software tools that automate the synthesis process that turns a high-level design of Boolean equation, state machines, and various high-level functions into a low-level design of simple cells from the library. They also support physical layout tools that turn that low level design into a circuit layout for a physical semiconductor chip. The ASIC vendor may even supply application engineers to assist the ASIC designer with the task of designing the chip and running the tools. The vendor then lays out the chip, creates the masks for the various layers, and manufactures the ASICs. Manufacturing the ASIC may be the most important function that the ASIC vendor provides. Semiconductor chip fabrication facilities ("fabs") cost on the order of $1 billion dollars to build these days. They require expensive equipment to handle silicon wafers and create the precise, sub-nanometer features that comprise the millions or billions of transistors on the chips. Companies other than the largest in the world can hardly afford to build a single fab, so the ASIC vendor uses their own fab, or sometimes uses the fabs of larger, established semiconductor manufacturing companies, to manufacture integrated circuits for many smaller companies.

The ASIC designer, who is a customer of the ASIC vendor, designs an integrated circuit at a high level of Boolean logic, state machines, and predesigned functions like processors and memories. Just as a board designer does not need to have an intimate knowledge of the integrated circuits that he places on a printed circuit board, the ASIC designer does not need such intimate knowledge of the individual cells that are used in an

ASIC design. This is not to say that no knowledge is required. Just as a PC board designer needs to know interface characteristics such as capacitive loading and trace impedance, an ASIC designer needs to understand the ASIC vendor's specifications for the particular library of cells that he is using in his design including what the inputs require, what the outputs will be, and something about the semiconductor process characteristics.

10.1 Gate Array vs. Standard Cell vs. Structured ASIC

There are three varieties of ASICs, and each has its own advantages - gate arrays, standard cells, and structured ASICs. Each variety has a different architecture as shown in Figure 74. These architectural differences result in different manufacturing techniques, different costs, and different development times.

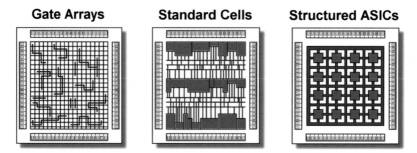

Figure 74. ASIC architectures

10.1.1 The Gate Array

The gate array consists of rows and columns of regular transistor structures. Each basic cell, or gate, consists of the same small number of transistors that are not connected. In fact, none of the transistors on the gate array are initially connected at all. The reason for this is that the connection is determined completely by the design that you implement. Once you have your design, the layout software figures out which transistors to connect. First, your low level functions are connected together. For example, six transistors could be connected to create a flip-flop. These six transistors would be located physically very close to each other. After your low level functions have been routed, these would in turn be connected together. The layout software would continue this process until the entire design is complete.

Figure 75 shows the basic parts of a gate array architecture. The gates are the unconnected transistors laid out in an array. The metal routing comprises the metal layers that are put onto the chip that converts it from a bunch of unconnected transistors to transistors connected in a specific way to implement your specific design. The input/output (I/O) cells get physically connected to the pins of the chip so that signals can be input to the chip and output from the chip.

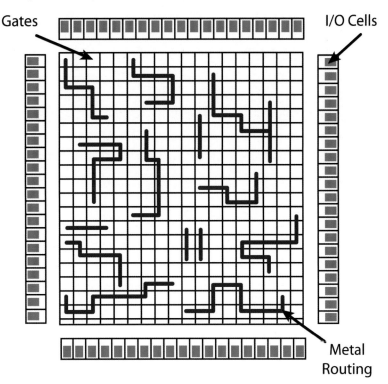

Figure 75. Gate array architecture

The ASIC vendor manufactures many unrouted die which contain the arrays of gates and which it can use for any gate array customer. An integrated circuit consists of many layers of materials including semiconductor material (e.g., silicon), insulators (e.g., oxides), and conductors (e.g., aluminum). An unrouted die is processed with all of the layers except for the final metal layers that connect the gates together. Once your design is complete, the vendor simply needs to add the last metal layers to the die to create your chip.

An advantage of gate arrays is that they have a fast turnaround time because they are almost complete before the design is even started. In addition, since the vendor can produce many unrouted arrays for many customers, each customer shares in some of the development cost, resulting in a lower development charge, also known as non-recurring expense (NRE).

10.1.2 The Standard Cell

The standard cell ASIC is designed using cells of transistors that are already connected together and compactly routed to form higher level functions such as flip-flops, adders, counters, multipliers, and even entire processors. The ASIC designer connects these cells together just as he would connect chips together on a PC board. The software that lays out a standard cell ASIC attempts to place these cells on the die and connect them together in as efficient a way as possible.

Figure 76 shows the basic parts of a standard cell architecture. The logic cells are the higher level functions consisting of several or many transistors that you specifically choose for your design. The metal routing comprises the metal layers that are put onto the chip that connects these logic cells in a specific way to implement your specific design. The input/output (I/O) cells get physically connected to the pins of the chip so that signals can be input to the chip and output from the chip.

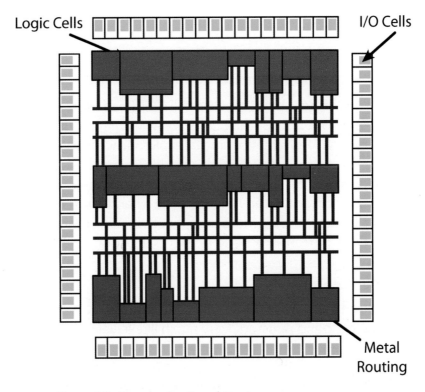

Figure 76. Standard cell architecture

Since each cell consists of all of the material layers needed to produce the transistors and to connect them, and since each customer's design is different, each standard cell ASIC must be created from scratch. This results in a much longer turnaround time than for a gate array or structured ASIC. Each mask to produce each layer is custom for each user. Therefore, customers cannot share development costs as they can with gate arrays and structured ASICs. This also results in a large non-recurring expense (NRE), which is a one-time charge that the customer pays to begin the standard cell ASIC design.

An advantage of a standard cell approach is that the resulting die is typically much smaller than the equivalent gate array or structured ASIC. For a gate array or structured ASIC, the die size is fixed and many transistors in the array will typically not be used in the design. For a standard cell design, only those transistors that are needed are placed on the die. A smaller die results in more die per wafer, which results in a smaller cost per part because there is a fixed cost to create a single wafer. This can be a big advantage for parts that are used in high volume.

Another advantage is that standard cell ASICs can use very complex functions if those functions are available as cells in the vendor's library. Many vendors include microprocessor cores in their libraries. These cells would be very difficult to design and would take up a great deal of die area if they were implemented in a gate array or structured ASIC because a hand-crafted custom cell will always be more compact than one created from a regular array of transistors and perpendicular connections.

10.1.3 The Structured ASIC

The structured consists of rows and columns of large cells containing logic and registers, very much like a Field Programmable Gate Array (FPGA), which is a kind of programmable chip that I will discuss in the next chapter. Each basic cell is identical. There are metal connections throughout the chip. Like a gate array, all of the cells are unconnected until additional layers of metal are added by the user to connect the cells into a specific function. Unlike a gate array, the logic blocks in a structured ASIC are at higher level than simple transistors and typically include logic gates and flip-flops. Figure 77 shows the basic pieces of the structured ASIC architecture including the macro blocks that contain logic components, the routing resources that are metal lines that must be connected appropriately to route connections from one logic block to another, and the input/output (I/O) cells that gets physically connected to the pins of the chip so that signals can be input to the chip and output from the chip.

As with gate arrays, the structured ASIC vendor manufactures many unrouted die which contain the arrays of cells and which it can use for any structured ASIC customer. The unrouted die is processed with all of the layers except for the final metal layers that connect the gates together. Once your design is complete, the vendor simply needs to add the last metal layers to the die to create your chip.

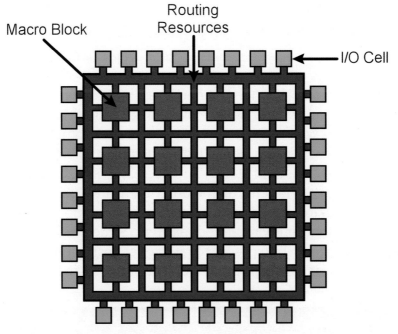

Figure 77. Structured ASIC architecture

In the past years as programmable chips, specifically FPGAs, became more popular, they displaced gate arrays in many designs. If a customer needed very quick turnaround time for a chip and was willing to put up with a high cost per chip, that customer would choose an FPGA. If the customer needed high performance and low cost per chip and had enough money to pay for the up-front NRE costs, that customer would choose a standard cell ASIC. ASIC vendors were losing money as ASIC designs decreased significantly each year. The solution that some ASIC vendors created is the structured ASIC that has the cost and turnaround time advantages of the gate array but has the architecture of the FPGA. Ironically, FPGAs were architected to be like gate arrays in order to take business away from gate array ASICs. A customer who was used to creating a gate array could easily move designs to an FPGA. Now, a structured ASIC has an architecture like an FPGA to try to win those designs back to the ASIC vendors. Whether this strategy works remains to be seen.

10.2 Gate Array vs. Standard Cell vs. Structured ASIC

Which ASIC type to use depends on your project and budget. Use gate arrays or structured ASICs when you want to hold down the initial cost, when you need fast turnaround on prototypes, and when you expect low production volumes. Use standard cells when you need to implement very complex functions and when you expect high production volumes. As FPGAs have increased in density and decreased in cost, their very fast turnaround time, on the order of minutes or hours, has allowed them to eat away a significant portion of the gate array ASIC market until it is virtually nonexistent. Structured ASICs have not done much to prevent that market destruction, and so practically, today's engineer has a choice between an FPGA and a standard cell ASIC.

	Gate Array	Standard Cell	Structured ASIC
NRE	low	high	low
Per Piece Cost	high	low	high
Utilization	low	high	low
Turnaround Time	fast	slow	fast

Table 2 ASIC Architecture Comparison

Quiz 9: ASICs

1. **What does the term ASIC stand for?**
   - ☐ Application Standard Integrated Cell
   - ☐ Apply Strenuously In California
   - ☐ Application Specific International Circuit
   - ☐ Application Specific Integrated Circuit

2. **Give the name of the ASIC architecture shown below.**

   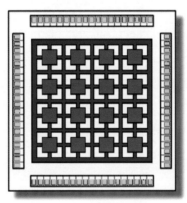

   - ☐ Structured ASIC   ☐ PROM
   - ☐ Standard Cell   ☐ Gate Array

3. **Which of the following things does an ASIC vendor not supply?**
   - ☐ Finished IC chips
   - ☐ The high-level chip design
   - ☐ Engineers to work on the chip
   - ☐ Physical layout tools

Application Specific Integrated Circuits (ASICs)

4. **Which of the following things does an ASIC designer need to know?**

    ☐ How to manufacture chips

    ☐ High-level chip design

    ☐ Silicon electrical characteristics

    ☐ Semiconductor process physics

5. **Give the name of the ASIC architecture shown below.**

    ☐ Structured ASIC   ☐ PROM

    ☐ Standard Cell   ☐ Gate Array

6. **Which of the following ASIC architectures consists of rows and columns of large cells containing logic and registers?**

    ☐ Gate array

    ☐ Standard cell

    ☐ Structured ASIC

7. **Which of the following ASIC architectures consists of rows and columns of regular transistor structures?**

    ☐ Gate array

    ☐ Standard cell

    ☐ Structured ASIC

8. Give the name of the ASIC architecture shown below.

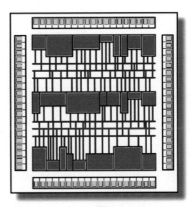

☐ Structured ASIC ☐ PROM

☐ Standard Cell ☐ Gate Array

9. Which of the following ASIC architectures is designed using cells of transistors that are already connected together and compactly routed to form higher level functions such as flip-flops, adders, counters, multipliers, and even entire processors?

☐ Gate array

☐ Standard cell

☐ Structured ASIC

10. Which of the following ASIC architectures is the only one that has so far survived in the market against Field Programmable Gate Array (FPGA) technology?

☐ Gate array

☐ Standard cell

☐ Structured ASIC

Application Specific Integrated Circuits (ASICs)

11. Select all of the advantages for each ASIC architecture in the table below.

	Gate Array			Standard Cell			Structured ASIC		
Initial cost (NRE)	a.	high	low	b.	high	low	c.	high	low
Per piece cost	d.	high	low	e.	high	low	f.	high	low
Utilization	g.	high	low	h.	high	low	i.	high	low
Turn around time	j.	fast	slow	k.	fast	slow	l.	fast	slow

12. Which of the following ASIC architectures is most similar to the architecture of a Field Programmable Gate Array (FPGA)?

    ☐ Gate array
    ☐ Standard cell
    ☐ Structured ASIC

11. Programmable Devices

Programmable devices are integrated circuits that a user can program, without any complicated or expensive machinery, to have the behavior required for the chip. An inexpensive, easy-to-use desktop device called a "device programmer" is used to load a series of bits into the device to program its behavior. Programmable devices have gone through a long evolution to reach the complexity that they have today. The following sections give an approximately chronological discussion of these devices from least complex to most complex.

11.1 Programmable Read Only Memories (PROMs)

As discussed in Section 9.2.1 Programmable Read Only Memories (PROMs), PROMs are simply memories that can be inexpensively programmed by the user to contain a specific pattern of ones and zeroes. The pattern remains even after power is removed from the device. This pattern can be used to represent a computer program for a microprocessor, an algorithm, or a state machine. Some PROMs can be programmed once only. Other PROMs, such as EPROMs or EEPROMs can be erased and programmed multiple times.

PROMs are excellent for implementing any kind of combinatorial logic (also called "Boolean logic") that requires a limited number of inputs and outputs. For sequential logic where steps need to be taken in a particular order, as with a state machine, external clocked devices such as flip-flops or microprocessors must be added. PROMs tend to be extremely slow, so they are not useful for applications where high speed is required.

11.2 Programmable Logic Arrays (PLAs)

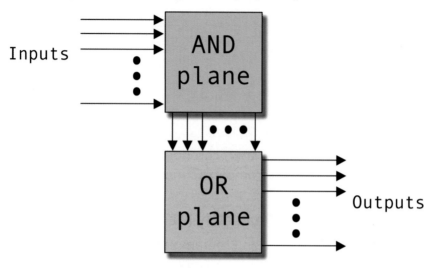

Figure 78. PLA Architecture

Programmable Logic Arrays (PLAs) were a solution to the speed and input limitations of PROMs. A PLA architecture is shown in Figure 78. PLAs consist of a large number of inputs connected to an AND plane, where different combinations of signals can be logically ANDed together. The specific signals that are ANDed together are determined by how the part is programmed. The outputs of the AND plane go into an OR plane, where the terms are ORed together in different combinations, also determined by how the part is programmed. Finally outputs are produced. At the inputs and outputs there are typically inverters so that a logical NOT of an input or output can be produced. These devices can implement a large number of combinatorial functions of the inputs, though not all combinatorial functions are possible as is the case with a PROM. However, they generally have many more inputs than a PROM and are much faster.

To give an illustration of the combinatorial logic outputs that a PLA can produce, suppose the inputs to the PLA are called signals A, B, and C while the output is called Z. Here are just a few examples of the outputs that this hypothetical, very simple PLA can output (remember from Section 7.1 Boolean Algebra that & is the symbol for AND, | is the symbol for OR, and ~ is the symbol for NOT).

Z = A & B
Z = A | C
Z = A & B & C
Z = ~(A & B) | ~C
Z = ~((A | B) & (~B | ~C) & (A | ~C))

11.3 Programmable Array Logic (PALs)

The Programmable Array Logic (PAL) is a variation of the PLA, with an architecture shown in Figure 79. Like the PLA, it has a wide, programmable AND plane for ANDing inputs together. However, the OR plane is small and fixed, limiting the number of terms that can be ORed together. In the figure, the AND plane is represented by the crisscrossing wires while the OR place is simply shown as OR gates with a fixed number of inputs.

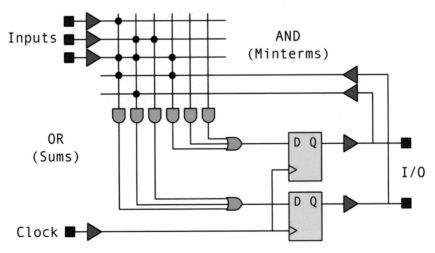

Figure 79. PAL Architecture

Other basic logic devices, such as multiplexers, exclusive ORs, and latches are added to the inputs and outputs. Most importantly, clocked elements, typically flip-flops, are included. These devices are now able to implement a large number of logic functions including clocked sequential logic needed to create state machines. This was an important development that allowed PALs to replace much of the standard logic in many designs. PALs are also extremely fast.

Ideally, though, the hardware designer wanted something that gave him or her the flexibility and complexity of an ASIC but with the shorter turn-around time of a programmable device. The solution came in the form of two new devices - the Complex Programmable Logic Device (CPLD) and the Field Programmable Gate Array. As can be seen in Figure 80, CPLDs and FPGAs bridge the gap between PALs and Gate Arrays. CPLDs are as fast as PALs but more complex. FPGAs approach the complexity of Gate Array ASICs but are still programmable.

11.4 CPLDs and FPGAs

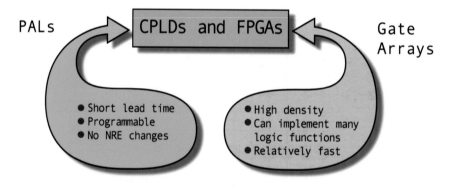

Figure 80. Comparison of CPLDs and FPGAs

11.5 Complex Programmable Logic Devices (CPLDs)

Complex Programmable Logic Devices (CPLDs) are exactly what they claim to be. Essentially they are designed to appear just like a large number of PALs in a single chip, connected to each other through a crosspoint switch. They require the same development tools and device programmers as PALs, and are based on the same technologies, but they can handle much more complex logic and more of it.

11.5.1 CPLD Architectures

The diagram in Figure 81 shows the internal architecture of a typical CPLD. While each manufacturer has a different variation, in general they are all similar in that they consist of function blocks (FBs), input/output blocks (I/O), and an interconnect matrix. The devices are programmed using programmable elements that, depending on the technology of the manufacturer, can be EPROM cells, EEPROM cells, or Flash EEPROM cells.

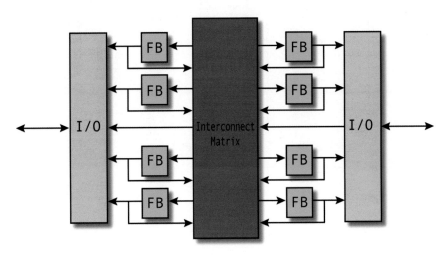

Figure 81. CPLD Architecture

11.5.1.1 Function Blocks

A typical function block is shown in Figure 82. The AND plane still exists as shown by the crossing wires. The AND plane can accept inputs from the I/O blocks, other function blocks, or feedback from the same function block. The terms are then ORed together, and terms are selected via a large multiplexer (or "mux"). The outputs of the mux can then be sent straight out of the block, or through a clocked flip-flop. This particular block includes additional logic such as a selectable exclusive OR and a master reset signal, in addition to being able to program the polarity of the signal (inverting or non-inverting) at different stages.

Usually, the function blocks are designed to be similar to existing PAL architectures, so that the designer can use familiar tools or even older designs without changing them, allowing them to easily migrate multiple legacy PAL designs into a single CPLD.

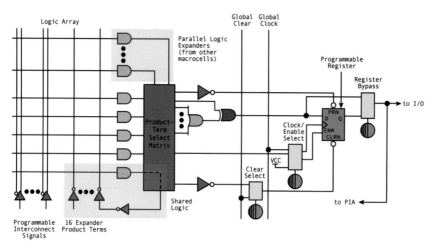

Figure 82. CPLD Function Block

11.5.1.2 I/O Blocks

Figure 83 shows a typical I/O block of a CPLD. The I/O block is used to drive signals to the pins of the CPLD device at the appropriate voltage levels with the appropriate current. Usually, a flip-flop is included, as shown in the figure. This is done on outputs so that clocked signals can be output directly to the pins without encountering significant delay. It is done for inputs so that there is not much delay on a signal before reaching a flip-flop. Also, some small amount of logic is included in the I/O block simply to add some more resources to the device.

11.5.1.3 Interconnect

The CPLD interconnect is a very large programmable switch matrix that allows signals from all parts of the device go to all other parts of the device. While no switch can connect all internal function blocks to all other function blocks, there is enough flexibility to allow many combinations of connections.

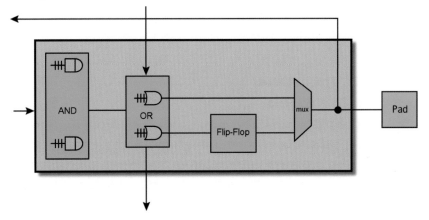

Figure 83. CPLD Input/Output Block

11.5.1.4 Programmable Elements

Different manufacturers use different technologies to implement the programmable elements of a CPLD. The common technologies are Erasable Programmable Read Only Memory (EPROM), Electrically Erasable PROM (EEPROM) and Flash EEPROM. These technologies are similar, or next generation versions of, the technologies that were used for the simplest programmable devices, PROMs.

11.6 Field Programmable Gate Arrays (FPGAs)

Field Programmable Gate Arrays are called this because rather than having a structure similar to a PAL or other programmable device, they are structured very much like a gate array ASIC. On their initial introduction to the market, this made FPGAs very nice for use in prototyping ASICs, or in places where an ASIC would eventually be used. For example, an FPGA could be used in a design that needed to get to market quickly regardless of cost. Later an ASIC could be used in place of the FPGA when the production volume increases, in order to reduce cost. However, as the density of FPGAs increased over the years, and their costs dropped dramatically as they were designed into products more frequently, FPGAs have actually destroyed much of the ASIC business except where very high volume or very high speed was a requirement that overrides their significantly higher initial expense and longer time to production.

11.6.1 FPGA Architectures

Each FPGA vendor has its own FPGA architecture, but in general terms they are all a variation of that shown in Figure 84. The architecture consists of configurable logic blocks, configurable I/O blocks, and programmable interconnect. Also, there is clock circuitry for driving the clock signals to each logic block and additional high functionality resources ("hard IP") such as ALUs, memory, and decoders may be available. The two basic types of programmable elements for an FPGA are Static RAM and antifuses.

Just Enough Electronics to Impress Your Friends and Colleagues

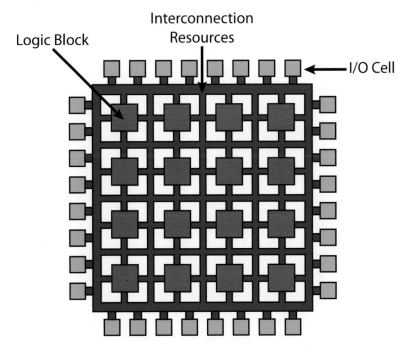

Figure 84. FPGA Architecture

11.6.1.1 Configurable Logic Blocks

Configurable Logic Blocks ("CLB") contain the logic for the FPGA. In a large grain architecture, these CLBs will contain enough logic to create a small state machine. In a fine grain architecture, more like a true gate array ASIC, the CLB will contain only very basic logic. The diagram in Figure 85 would be considered a large grain block and it is by far the most common FPGA architecture. It contains RAM for creating arbitrary combinatorial logic functions. It also contains flip-flops for clocked storage elements, and multiplexers in order to route the logic within the block and to and from external resources. The muxes also allow the selection of the polarity of signals (inverted and non-inverted signals) as well as selection of which signals are used to set (SD input) and reset (RD input) the flip-flop outputs.

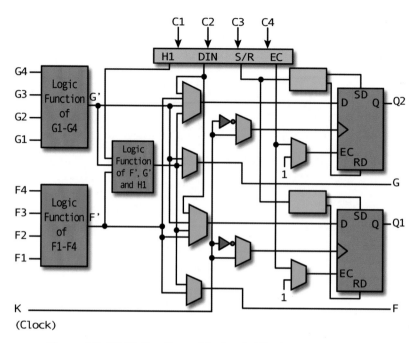

Figure 85. FPGA Configurable Logic Block

11.6.1.2 Configurable I/O Blocks

A Configurable I/O Block, shown in Figure 86, is used to bring signals onto the chip and send them back off again. It consists of an input buffer and an output buffer with tri-state output controls, which means that the outputs can be a high voltage signal (a logic 1), a low voltage signal (a logic 0), or disconnected from the circuit (thus having three states rather than simply two binary states). The polarity of the output can usually be programmed. Often the slew rate (the speed at which the signal changes voltage) of the output can be programmed for fast changes to speed up the circuit or slow changes to minimize high frequency electronic noise that can disturb other nearby devices. In addition, there is often a flip-flop on outputs so that clocked signals can be output directly to the pins without encountering significant delay. Flip-flops are often also used in I/O blocks for inputs so that there is not much delay on a signal coming into the chip before reaching a flip-flop.

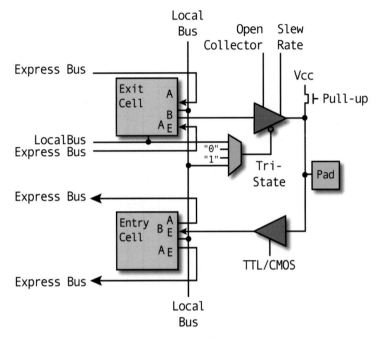

Figure 86. FPGA Configurable I/O Block

11.6.1.3 Programmable Interconnect

The interconnect of an FPGA is very different than that of a CPLD, but is rather similar to that of a gate array ASIC. In Figure 87, a hierarchy of interconnect resources can be seen. There are long lines that can be used to connect critical CLBs that are physically far from each other on the chip without introducing much delay, because signals are delayed every time they must go through a transistor that connects one interconnect line to another. There are also short lines that are used to connect individual CLBs that are located physically close to each other. There are often one or several switch matrices, like that in a CPLD, to connect these long and short lines together in specific ways. Programmable switches inside the chip allow the connection of CLBs to interconnect lines and the connection of interconnect lines to each other and the connection of CPBs and interconnect lines to the switch matrix. Tri-state buffers are used to connect many CLBs to a long line, creating a "bus" where different CLBs can be connected and disconnected from the bus as required during normal operation.

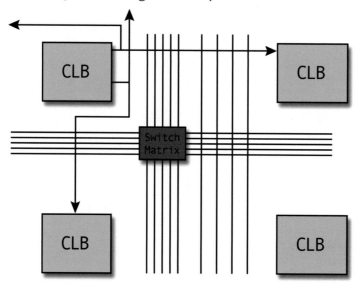

Figure 87. FPGA Programmable Interconnect

11.6.1.4 Clock Circuitry

Special I/O blocks with special buffers that can drive signals very quickly, known as clock drivers, are distributed around the chip. These buffers are connected to clock input pads and drive the clock signals that are input to the chip onto specific long interconnect lines, called global clock lines. These global clock lines are specially designed for fast propagation times so that clock signals get to each part of the chip at nearly the same exact time. These clock lines are connected to each clocked element in each CLB.

11.6.1.5 SRAM vs. Antifuse vs. Flash Programming

There are three competing methods of programming FPGAs. SRAM programming uses small static RAM bits for each programming element. Writing the bit with a zero turns off a switch, while writing with a one turns on a switch. Another method involves antifuses, which consist of microscopic structures that, unlike a regular fuse, normally make no connection. A certain amount of current during programming of the device causes the two sides of the antifuse to connect. A third method uses Flash EEPROM bits for each programming element.

The advantages of SRAM based FPGAs is that they use a standard fabrication process that chip fabrication plants are familiar with and are always optimizing for better performance. Since the SRAMs are reprogrammable, the FPGAs can be reprogrammed any number of times, even while they are in the system, just like writing to a normal SRAM. SRAM based devices can easily use the internal SRAMs as small memories in the design if required. The disadvantages are that SRAMs are volatile, which means a power glitch could potentially change it and it must be reprogrammed every time the FPGA is powered up. Also, SRAM-based devices have large routing delays, making them generally slower in theory. In practice, however, SRAM based FPGAs are very fast because every semiconductor company can produce this standard technology and thus every semiconductor company is continually improving the SRAM process.

The advantages of antifuse based FPGAs are that they are non-volatile and the delays due to routing are very small, so they tend to be faster. Antifuse based FPGAs tend to require lower power and they are better for keeping your design information out of the hands of competitors because they do not require an external device to program them upon power-up as SRAM based devices do. The disadvantages are that they require a complex fabrication process, they require an external programmer to program them, and once they are programmed, they cannot be changed.

Flash-based FPGAs seem to combine the best of both of the other methods. They are nonvolatile like antifuse FPGAs, yet reprogrammable like SRAM-based FPGAs. They use a standard fabrication process like SRAM-based FPGAs and they are lower power and secure like antifuse FPGAs. They are also relatively fast.

Quiz 10: Programmable Devices

1. **What does the term PROM mean?**
   - ☐ Programmable Read Only Memory
   - ☐ Partial Read Only Memory
   - ☐ Programmable Random Only Memory
   - ☐ Poor Richard's Occupation Manual

2. **What does the term PLA mean?**
   - ☐ Proletariat Liberation Army
   - ☐ Programmable Logic Array
   - ☐ Partial Lithography
   - ☐ Programmable Little ASIC

3. **A PLA contains (check all that apply)**
   - ☐ A large programmable AND plane
   - ☐ A large programmable OR plane
   - ☐ A large programmable XOR plane
   - ☐ Inverters on the inputs and outputs
   - ☐ Flip-flops
   - ☐ A small fixed OR plane

4. **What does the term PAL mean?**
   - ☐ Programmable Array Logic
   - ☐ Partial Array Logic
   - ☐ Promise Anything Lightly
   - ☐ Programmable ASIC Lite

5. **A PAL contains (check all that apply)**
   - ☐ A large programmable AND plane
   - ☐ A large programmable OR plane
   - ☐ A large programmable XOR plane
   - ☐ Inverters on the inputs and outputs
   - ☐ Flip-flops
   - ☐ A small fixed OR plane

6. **What does the term CPLD mean?**
   - ☐ Common People Love Dancing
   - ☐ Complex Programmable Little Diode
   - ☐ CMOS Programmable Logic Device
   - ☐ Complex Programmable Logic Device

7. **A CPLD contains (check all that apply)**
   - ☐ A single large programmable switch matrix
   - ☐ Combinatorial logic blocks containing small RAMs, flip-flops, and muxes
   - ☐ I/O blocks
   - ☐ Programmable elements
   - ☐ Clock buffers and clock lines
   - ☐ Function blocks like individual PALs
   - ☐ A hierarchy of interconnect lines

8. What does the term FPGA mean?
    - ☐ Field Programmable Gate Array
    - ☐ Foul People Get Angry
    - ☐ Formal Programmable Giant ASIC
    - ☐ Field Programmable Giant ASIC

9. An FPGA contains (check all that apply)
    - ☐ A single large programmable switch matrix
    - ☐ Combinatorial logic blocks containing small RAMs, flip-flops, and muxes
    - ☐ I/O blocks
    - ☐ Programmable elements
    - ☐ Clock buffers and clock lines
    - ☐ Function blocks like individual PALs
    - ☐ A hierarchy of interconnect lines

10. Which technology is not used in a CPLD for programming?
    - ☐ EEPROM
    - ☐ SRAM
    - ☐ Flash EEPROM
    - ☐ EPROM

11. Which technology is not used in a FPGA for programming?
    - ☐ antifuses
    - ☐ SRAM
    - ☐ Flash EEPROM
    - ☐ DRAM

12. Match each programmable device with its description.

a)	PROM	A memory device that can be programmed once and read many times.
b)	PLA	A logic device that can be used to design large functions like an ASIC except that it can be programmed quickly and inexpensively.
c)	PAL	A logic device that is made up of many PAL devices.
d)	CPLD	A logic device with a large AND plane and a large OR plane for implementing different combinations of Boolean logic.
e)	FPGA	A logic device with a large AND plane and a small, fixed number of OR gates for implementing Boolean logic and state machines.

12. Computer Architecture

This section describes the architecture of a computer. We will look at the Von Neumann architecture, specifically, named after John Von Neumann, one of the inventors of the digital computer. This architecture, with various modifications, is the basis for almost all digital computers in existence.

12.1 System Architecture

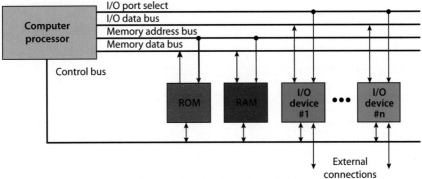

Figure 88. General Computer System Architecture

The architecture shown in Figure 88 is a generalized view of a computer. The processor contains all of the brains of the computer. If the processor is on an integrated circuit, it is known as a microprocessor. The processor executes instructions in the form of binary data. Collections of instructions that perform specific tasks are called programs. Programs that do not change are stored in Read-Only Memory (ROM), which is programmed at the factory and cannot be changed. These programs usually control the hardware directly. For example, the program may instruct the processor to access a specific input/output (I/O) device such as the keyboard or hard drive. In a personal computer (PC), these programs are called the Basic Input/Output System (BIOS).

In order to execute a program, the processor puts out a binary address on its memory address bus. A bus is simply a bunch of signals that are related in a specific way. In this case, each signal represents a binary number. That binary number is a location in memory from which the program wants to "fetch" an instruction. The memory responds by putting the contents of that location on the data bus. The Random Access Memory (RAM), which can be written and read by the processor, is used to store data. Note that in this architecture, the processor gets data in the same way it gets instructions. In some processors, using what is called a Harvard architecture, there is a separate data bus and instruction bus. This allows the processor to output instruction addresses at the same time as data addresses, speeding up program execution. This also allows some other optimizations to be made since data patterns in memory are different from instruction patterns.

Data can also be communicated to an input/output (I/O) device such as a hard drive, computer monitor, network connection, keyboard, etc. These devices are accessed through I/O port selects, which are signals that enable and disable the I/O device. The data to and from an I/O device go through an I/O data bus. The data can actually be instructions, such as program instructions stored on a hard drive, or they can be information, such as files stored on a hard drive or user input from a keyboard or mouse. The I/O devices may have external connections to other devices, such as a universal serial bus (USB) that may be connected to hard drives, scanners, CD-ROMs, etc.

The internal architecture of a processor is shown in Figure 89. The diagram is broken into two sections, a data section and a control section. The data section handles the flow of data through the processor, including any operations that take place using the data or in some way change the data. The control section handles execution of the instructions that control how the data is manipulated.

12.2 Processor Architecture

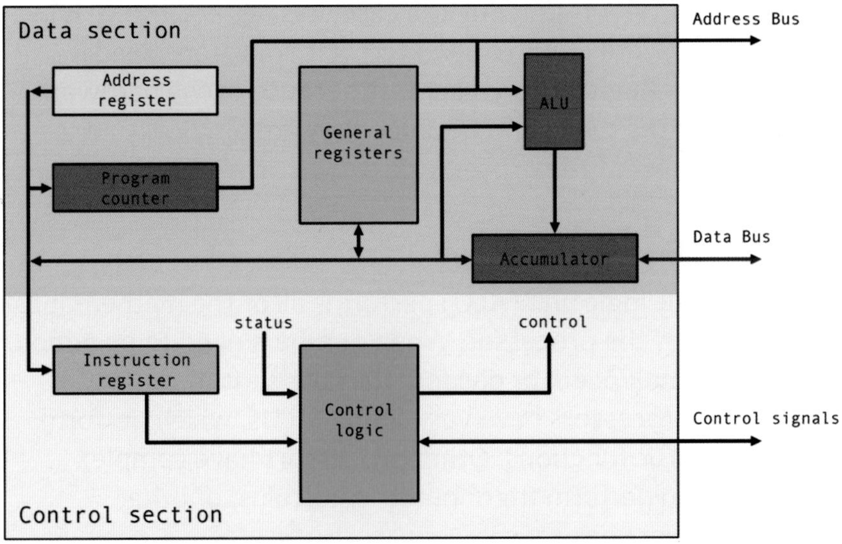

Figure 89. Processor Internal Architecture

12.2.1 Address Register

The address register is simply used to hold the address for a data access of memory. Addresses can be manipulated by the processor just like other data. The address is put out on the address bus in order to access memory.

12.2.2 Program Counter

The program counter is similar to the address register, except that it is used as an address for the next program instruction to fetch from memory. It can also be manipulated, but there are typically limits, since program instructions are usually accessed in sequential order and near each other. Data, on the other hand, can be located anywhere in memory and the location of one byte of data may have no relationship to the location of the next byte of data.

12.2.3 General Registers

The general registers are simply locations in the processor for temporarily holding data. When data needs to be accessed several times, as is frequently the case, it is faster to store it in a register than write it out to an external memory and read it back in again. Registers are very fast because of the hardware that implements them, and because they are physically very close to the rest of the processor.

12.2.4 Arithmetic Logic Unit (ALU)

The arithmetic logic unit (ALU) performs all of the mathematical calculations of the processor. When numbers need to be added, subtracted, multiplied, or divided, they are sent through the ALU. Some processors have very simple ALUs, which perform only these basic functions. Other processors have complex ALUs that can perform trigonometric, calculus, or other specialized functions.

12.2.5 Accumulator

The accumulator is the special register that holds the results from an ALU operation. After a calculation is specified, an instruction in the program must tell the processor to retrieve the result from the accumulator and place it in a register or memory.

12.2.6 Instruction Register

The instruction register is the register that holds the current instruction in the program. The program counter points to a location in memory where this instruction is located. The processor fetches that instruction and places it in the instruction register where it decodes it, figuring out what steps it needs to take to execute it.

12.2.7 Control Logic

The control logic is a state machine in the processor that figures everything out and controls the rest of the hardware in order to execute that instruction. Every piece of hardware described so far has a very straightforward function. Registers simply hold data. Even the ALU takes the data given to it, performs the specified operation, and stores the result in the accumulator. It is the control logic that, based on the instruction it fetches, controls when a register will store data, when data is to be written to memory, or when the ALU will perform an operation and which operation it should perform.

12.3 Pipelined Processors

Most processors today are pipelined processors. Pipelined refers to the control logic of the processor. It means that operations takes place as if in an assembly line, where each piece of hardware performs a small, specific function before sending the result downstream to the next piece of hardware in the line. In this way, each piece of hardware is working at the same time, rather than waiting for its turn, making the processor execute instructions much faster than a non-pipelined processor.

A pipelined processor may, for example, have four stages as shown in Figure 90 - instruction fetch, decode, execute, and load/store - although there are many variations. The instruction fetch stage fetches the most likely next instruction. It may turn out that this instruction is not needed because the program will go off to another section of the program based on incoming data or the result of a calculation. This is called a program branch. However, the processor retrieves what it believes is most likely to be the next instruction. If, at any time before executing it, the processor determines that this is not the next instruction to execute, the processor will "toss out" this instruction and fetch the correct one.

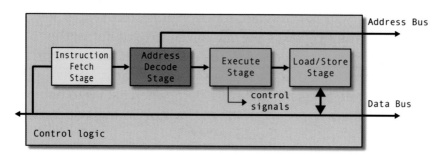

Figure 90. Processor Pipeline

The decode stage decodes the instruction and plans the operations that must be performed according to the instruction. If a location in memory is to be accessed, the address is set up during this stage and written to the address register.

The execute stage actually executes the plan determined from the decode stage, controlling the ALU and any other hardware that is necessary.

The load/store stage handles any accesses of memory. Reading and writing data to and from memory is often a very slow process compared to the other stages. By pipelining, the memory can be accessed while the processor continues executing other instructions as long as they don't require the data being accessed by the load/store stage.

Many of today's processors are pipelined and include redundant hardware. This allows multiple instructions to be executed at the same time, greatly speeding up program run times. For example, suppose we have the following instructions in a program:

```
LOAD VALUE FROM ADDRESS 15 IN MEMORY
ADD 5
STORE NEW VALUE TO ADDRESS 36 IN MEMORY
```

The diagram in Figure 91 compares how this program would be executed by a non-pipelined computer and a pipelined one. The speedup due to pipelining is apparent.

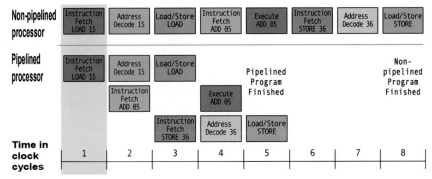

Figure 91. Pipelined vs. Non-pipelined Performance

12.4 Cache Memory

Cache memory is a very small but fast memory that resides between the processor and the main memory. When data is read into the processor, a copy of the data is stored in the cache. The next time the processor needs to read the same data, it gets it from the fast cache, rather than the slow main memory. When the processor needs to access data many times in a row, each access after the first will be very fast, significantly speeding up execution of the program. Since most programs contain loops - sets of instructions that are executed many times in a row - caches are particularly useful for instructions. Some data must also be accessed many times, and caches can help for these data accesses also. When the processor writes data into memory, the cache is also updated. A simplified cache memory architecture is shown in Figure 92.

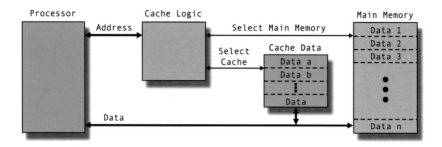

Figure 92. Cache Memory

12.5 RISC vs. CISC

The first computers that were designed, and many computers today, have an architecture that has come to be known as CISC (pronounced "sisk") for "complex instruction set computer." This architecture is based on the idea that complex programs require complex instructions. So very complex instructions were created to do very long series of operations. This makes the control logic section of the processor very big, very complex, and somewhat slow. One problem that researchers saw is that most programs use only the simple instructions, although they may use them in very complex combinations. By making the control logic complex and slow, it slowed down the entire operation. A new architecture, RISC (pronounced "risk"), for "reduced Instruction set computer," essentially removed all of the complex operations that were not used or that could be performed by a long sequence of simple operations. This reduced the complexity of the control logic, making it operate faster. This also sped up the development time for designing the computer and reduced the number of bugs in the hardware. Professors David Patterson and Carlo Sequin at the University of California at Berkeley and Professor John Hennessey at Stanford University invented and championed RISC processor design in 1980.

Most complex processors today are called RISC processors, but in actuality they combine many aspects of RISC and CISC design concepts.

12.6 Very Long Instruction Word (VLIW) Computers

An alternative computer architecture is the very long instruction word (VLIW) computer, pioneered by Professor Josh Fisher at Yale University in the early 1980s. This architecture was abandoned until recently when it has gained new supporters. The antithesis of RISC architecture, a VLIW computer has, as the name says, a very long instruction consisting of many bits. A very long instruction allows many operations to take place simultaneously as long as the processor has hardware resources available.

Computer Architecture

Quiz 11: Processors

1. The man who invented the modern computer architecture was?

    ☐ Werner Von Braun

    ☐ John Von Neumann

    ☐ Eddie Van Halen

2. The main parts of a general purpose computer are (check all that apply)

    ☐ ROM   ☐ LEDs   ☐ processor

    ☐ RAM   ☐ engine ☐ I/O devices

3. Which are not buses in a general purpose computer architecture?

    ☐ memory data bus   ☐ school bus

    ☐ control bus   ☐ I/O data bus

    ☐ universal serial bus ☐ Harvard bus

4. The two main sections of a computer processor are

    ☐ data section

    ☐ address section

    ☐ control section

5. Check all of the functions that can be found in the data section of a computer processor.

    ☐ ALU   ☐ address register

    ☐ accumulator   ☐ program counter

    ☐ general registers ☐ instruction register

6. Check all of the functions below that can be found in the control section of a computer processor.

    ☐ radiator
    ☐ control logic
    ☐ address register
    ☐ discombobulator
    ☐ instruction register
    ☐ accumulator

7. What does ALU mean?

    ☐ always look up
    ☐ arithmetic logic unit
    ☐ algebraic logic unit

8. Select all of the following that can be stages of a pipelined processor.

    ☐ load/store
    ☐ execute
    ☐ address decode
    ☐ instruction fetch

9. The small, fast memory that is used to speed up accesses to the main memory is called a

    ☐ write only memory
    ☐ read only memory
    ☐ cache memory

10. What does the term RISC mean?

    ☐ reduced instruction set computer
    ☐ remember it says computer
    ☐ relevant internal system computer

Just Enough Electronics to Impress Your Friends and Colleagues

11. What does the term CISC mean?

☐ combined instruction set calculator

☐ complex instruction set computer

☐ combinatorial instruction set computer

12. What does the term VLIW mean?

☐ very little internet waiting

☐ view less in water

☐ very long instruction word

Computer Architecture

13. Engineering Equipment

An important, and practical, part of understanding electronics is understanding the equipment that is used to design, debug, and manufacture hardware. The following sections describe some of the more basic tools that electrical engineers use.

Figure 93 shows a digital multimeter (DMM) that is a combination of several very important electronics meters. It is a hand-held device with two probes and a dial to change the operation and the sensitivity. The probes are placed on two different points of a circuit to make measurements.

13.1 Voltmeter

A voltmeter can be a standalone device but is often one of several meters that can be found in a digital multimeter. When the DMM is placed in voltmeter operation, the device measures the voltage difference between the two probe points, in volts. A DC voltmeter is intended to measure direct current voltages that are constant. An AC voltmeter is intended to measure alternating current voltages that are changing on a periodic basis. When measuring an AC voltage, the voltmeter will give an RMS value, which is a kind of average value for the voltage difference between the two points. It is not a pure average because when a signal is high voltage for the same amount of time that it is a low voltage, the average is zero, which is not too helpful as a measure of AC voltage.

Engineering Equipment

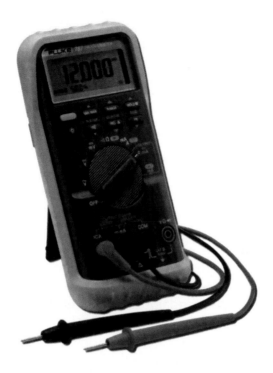

Figure 93. A digital multimeter (DMM)

13.2 Ohmmeter

An ohmmeter can be a standalone device but is often one of several meters that can be found in a digital multimeter. When the DMM is placed in ohmmeter operation, the device measures the resistance between the two probe points, in ohms. A trace that connects two points and does not have any imperfections will have a resistance close to zero. An open trace, one that has a break in it, will have a very high resistance, essentially off the scale. When measuring the resistance of a semiconductor device, the device should be removed from the board first, since the other devices connected to it on the board will change the overall resistance measurement.

13.3 Ammeter

An ammeter can be a standalone device but is often one of

Just Enough Electronics to Impress Your Friends and Colleagues

several meters that can be found in a digital multimeter. A connection in a circuit is cut and the probes are placed at each end, so that the broken circuit is reconnected through the ammeter. The electric current of the circuit then has to go through the ammeter, which measures the amount of current going through the circuit.

13.4 Oscilloscope

An oscilloscope, like the one shown in Figure 94 is used to look at voltage signals at points on a circuit as they change in time. The oscilloscope has multiple probes that are placed at points in the circuit to be observed. While the circuit is running, the voltage at these points will change.

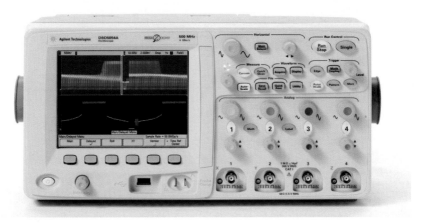

Figure 94. Oscilloscope

The voltage changes can be seen on the screen of the oscilloscope, as illustrated in Figure 95. Each horizontal line represents the voltage at a single point in the circuit. The screen shows how the signal changes in time.

Engineering Equipment

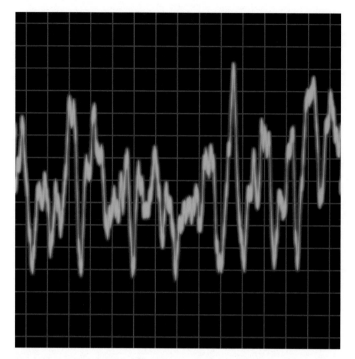

Figure 95. Oscilloscope screen

An analog oscilloscope sweeps an electron beam across the screen at a selected rate. The electron beam excites the phosphors on the screen, causing them to glow. When the voltage increases, the beam is driven up. When the voltage decreases, the beam is driven down. The glow of the phosphors persists for a short time, allowing the entire trace to be seen on the screen. An analog oscilloscope is useful for watching periodic signals such as clocks or audio signals.

A digital oscilloscope samples the incoming voltage signal at short, regular intervals and displays the waveform on the screen as a series of disconnected or connected dots. The advantage of a digital scope is that the samples are stored in memory. This allows a single pulse or glitch to be captured and displayed for examination even if that glitch only occurs once a minute, once an hour, or once without ever repeating. A digital scope is useful for finding unwanted glitches or intermittent problems with a signal as well as periodic signals. Analog storage scopes that have very long persistent phosphors also allow one-time events to be captured. However, with the advent of digital scopes, analog storage scopes are no longer manufactured.

13.5 Signal Generator

A signal generator, like that shown in Figure 96 is used to produce very specific types of signals that are input to a circuit for testing. For example, a signal generator may produce all of the various types of computer monitor signals for input to a prototype computer monitor. The computer monitor can then be tested to insure that it can correctly decode all of the formats that it is intended to display. The signal generator can also vary the signals that it produces within the extremes of the specification. In this way, the device under test (DUT) can be examined to make certain that it can handle all formats of input signals under all conditions.

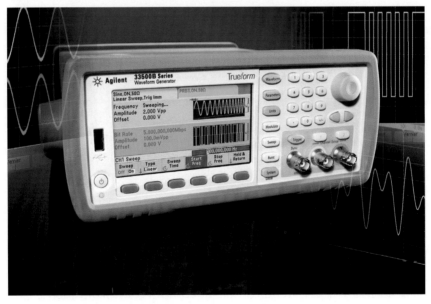

Figure 96. Signal generator

Engineering Equipment

13.6 Logic Probe

A logic probe is a simple device that looks like a fat pen with a metal tip and two wires coming out of the top, seen in Figure 97. It is used to examine digital systems. One wire is connected to the voltage representing a logic zero of the system (typically ground) while the other wire is connected to the voltage representing the logic one of the system (typically a positive voltage). The metal tip is then placed anywhere on the circuit. The logic probe will light an LED on its side to show whether that node is at a logic zero voltage or logic one voltage. The probe can be set up to show whether the node is changing between levels, whether there is a momentary pulse to a different level, or whether the node is turned off (high impedance) and not even connected to the circuit.

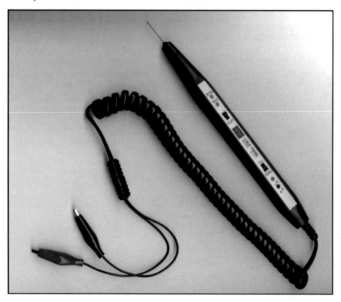

Figure 97. Logic probe

A logic probe is a very useful device for quickly examining digital circuits without the need for large, cumbersome, expensive equipment. Unfortunately, most engineers these days have not heard of logic probes and they are hardly ever used. This is sad because it has been a useful, inexpensive, easy-to-use, lightweight tool for many engineers debugging digital circuits.

13.7 Logic Analyzer

A logic analyzer, shown in Figure 98, is used to examine complex digital circuits. The wires that come from a logic analyzer are connected to nodes in the circuit. The logic analyzer can then be set up to look for very specific and very complex sequences of changes in the states of these nodes.

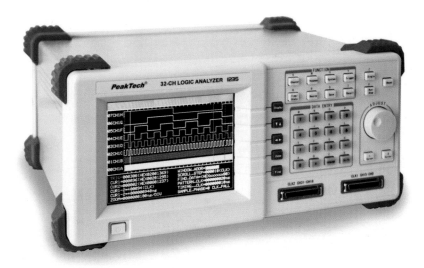

Figure 98. Logic analyzer

The user creates what is called a trigger, which is an event that occurs when a specific sequence of events has occurred. The logic analyzer can store the logic levels on each node at regular intervals of time. Each interval when a signal is stored is called a sample. The logic analyzer can be set up to record all samples before the trigger point, all samples after the trigger point, or a certain number of samples before and after the trigger point. Samples can be filtered so that only specific types of samples are stored. A logic analyzer allows the engineer to get very specific, detailed information about the operation of particular sections of very complex digital circuits.

Engineering Equipment

13.8 Protocol Analyzer

A protocol analyzer, like the one shown in Figure 99, is another piece of equipment intended for examining digital circuits. A protocol analyzer is connected between the communication ports of two or more digital systems. For example, one cable may be connected to a network wall socket while another cable is connected to the network port on a computer. All network signals would go through the protocol analyzer that would record them and allow an engineer to analyze the signal to find problems. The protocol can also be connected wirelessly to grab wireless network signals from the air and record them.

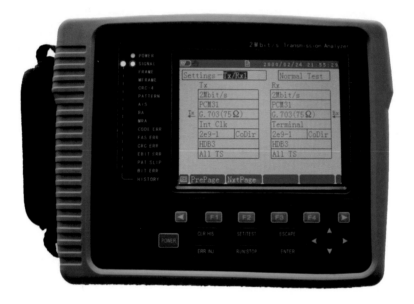

Figure 99. Protocol analyzer

The protocol analyzer stores the signals that are being transmitted over digital communication channels such as a network, decodes them, and interprets them. For example, the protocol analyzer may recognize that a certain series of ones and zeroes represents a special kind of data called an Ethernet packet. Or it may recognize that a certain series of packets is a digitization of voice or an e-commerce transaction. The protocol analyzer then displays the information in a form that a human can read and understand. In this way, complex communication data made up of many signals changing in a complex manner at high speeds can be captured and examined.

13.9 In-Circuit Emulator

An in-circuit emulator, or ICE, is shown in Figure 100. It is used to replace the microprocessor in a circuit while the circuit is being tested. The ICE can perform all of the complex functions that the actual microprocessor can do. In addition, the ICE records all of the microprocessor's activity so that a history of the chip can be obtained. The ICE can also be programmed to stop when a particular event occurs. This is useful if a bug occurs in a system only after a very long sequence of operations has occurred. For example, perhaps the computer monitor intermittently goes blank. The ICE can be programmed to stop operation when the signal to the monitor stops. When the ICE stops, the internal workings of the microprocessor can be examined along with a history of its operation. The system to which the ICE is connected can also be stopped and examined to find exactly which part of it failed.

When an integrated circuit manufacturer creates a new microprocessor, computer manufacturers want to ensure that their computer that uses the microprocessor will work correctly. They will want to make certain that all of the hardware, such as the monitor, keyboard, hard drive, and network connection, work correctly with the microprocessor. They also want to make sure that the computer can run the operating systems that are required for the computer. To debug all of the hardware and software in their new computer, they often use an ICE.

Engineering Equipment

Figure 100. In-circuit emulator (ICE)

Quiz 12: Equipment

1. Which three of the following devices might be found in a digital multimeter?

    ☐ voltmeter ☐ gas meter

    ☐ ammeter ☐ ohmmeter

    ☐ thermometer ☐ kilometer

2. What does a voltmeter measure?

    ☐ current ☐ voltage ☐ inductance

3. What does an ammeter measure?

    ☐ current ☐ resistance ☐ inductance

4. What does an ohmmeter measure?

    ☐ resistance ☐ capacitance ☐ inductance

5. The electrocardiogram (ECG) heart monitor in hospitals, which shows a line for the patient's heartbeat, is essentially

    ☐ an ammeter ☐ a protocol analyzer

    ☐ a logic analyzer ☐ an oscilloscope

6. The term "ICE" stands for

    ☐ in-circuit examiner

    ☐ in-circuit emulator

    ☐ internal circuit electronics

    ☐ internal cooling element

7. **Match the equipment to its function.**

a) Oscilloscope	Stores digital signals that are transmitted through it, decodes them, and interprets them.
b) Signal generator	Used to examine complex digital signals by storing the signal levels at specific times.
c) Logic probe	Used to look at voltage signals at points on a circuit, as the signals change in time.
d) Logic analyzer	A pen-like device that is used to examine digital systems.
e) Protocol analyzer	Used to replace a microprocessor in a real circuit for testing the circuit.
f) In-circuit emulator	Used to produce very specific types of signals that are input to a circuit for testing.

14. Resources

This is a list of resources that cover many of the preceding topics. Some of the books are written for non-engineers, while others are introductory engineering books. They can be used for reference or further study.

1. *Bebop to the Boolean Boogie: An Unconventional Guide to Electronics Fundamentals, Components, and Processes*, Clive "Max" Maxfield, LLH Technology Publishing, 1995.

2. *Bebop Bytes Back: An Unconventional Guide to Computers*, Clive "Max" Maxfield and Alvin Brown, Doone Publications, 1997. Another fun, off-beat book that focuses specifically on computers and computer technology.

3. *Basic Electronics*, Bernard Grob, Macmillan/McGraw-Hill, 1992. A beginner's guide to electronics.

4. *Designing with FPGAs & CPLDs*, Bob Zeidman, CRC Press, 2002. A good book for managers who need an overview or engineers needing background and design guidelines.

5. *Electronics the Easy Way*, Rex Miller and Mark R. Miller, Barron's Educational Series, 1984. A beginner's guide to electronics.

6. *How to Test Almost Anything Electronic*, Delton T. Horn, McGraw-Hill, 1993. A guide to testing common electronic devices.

7. *McGraw-Hill Electronics Dictionary*, Neil Sclater and John Markus, McGraw-Hill, 1997. A comprehensive explanation of the terminology of electronics.

8. *Teach Yourself Electricity and Electronics*, Stan Gibilisco, McGraw-Hill, 1997. A beginner's guide to electronics.

Index

15. Index

A

AC *See* alternating current
address bus 139, 141
alternating current 31, 153
ALU .... *See* arithmetic logic unit
American Bosch Arma Corporation .. 97
American Telephone & Telegraph
    Company 49
ammeter 154
amp 6, 24, 25
Ampère, André Marie 6
annealing 88
antifuse 97, 128, 133
application specific integrated circuit 2,
    3, 109, 110, 111, 112, 113, 114,
    115, 116, 124, 127, 128, 129, 131
    gate array . 110, 111, 112, 113, 114,
        115, 116, 124, 127, 129, 131
    standard cell 110, 112, 113, 114,
        115, 116
    structured ASIC . 110, 113, 114, 115,
        116
ASIC.*See* application specific integrated
    circuit
atom
    electron . 5, 6, 7, 8, 9, 11, 17, 24, 31,
        36, 58, 79, 81, 83, 88, 98, 99,
        104
    neutron 5, 79
    nucleus 5, 6, 79
    proton 5, 6, 11, 79

B

Bardeen, John 49
battery 8, 9, 31, 82, 83, 101
Bell Telephone Company 49
Bell, Alexander Graham 35, 49
binary 68, 69, 70, 130, 139
Boole, George 65
Boolean algebra .. 65, 66, 67, 68, 69, 70,
    71, 109, 121
Boolean logic .... *See* Boolean algebra
boule 87
Brattain, Walter 49

C

cache memory 145
capacitor . 23, 25, 26, 27, 28, 36, 38, 39,
    40, 41, 57, 98, 101, 102
chemical deposition 88
chemical etching 88
Chow, Wen Tsing 97
chronosynclastic infundibulum 1
CISC .... *See* complex instruction set
    computer
CLB .... *See* configurable logic block
clock lines 132
complex instruction set computer .. 146
complex programmable logic device . 3,
    124, 125, 126, 127, 131
conductor 35, 57, 79
conductors 6, 7, 17, 58, 79, 80, 112
configurable logic block 129, 132
CPLD. *See* complex programmable logic
    device
Czochralski method 86
Czochralski, Jan 86

D

data bus 140
DC *See* direct current
decimal 68, 69
delay element 70
Dennard, Robert 101
device under test 157
digital multimeter 153, 154, 155
diode 47, 49, 82, 83, 97
direct current 31, 153
dopants 88
DUT .... *See* device under test
dynamic random access memory 95,

100, 101, 102, 103, 104, 105, *See* dynamic random access memory

E

Edison, Thomas.................................. 31
Einstein, Albert 17
electric generator 11
    active solar 17
    fossil fuels....................................... 13
    geothermal..................................... 13
    nuclear.. 15
    passive solar 15
    steam... 13
    water 11
    wind... 16
electrically erasable read only memory 95, 98, 100, 101, 121, 125, 127, 133
electricity 5, 6, 9, 11, 13, 15, 16, 17, 23, 24, 25, 28, 31, 35, 36, 38, 47
electrode 8
    anode 8, 9, 31, 47, 82, 83
    cathode 8, 9, 31, 47, 82, 83, 88
electrolyte... 8
electromagnet 35, 36
electromagnetic force 5
electromagnetism............................. 11
EPROM.. *See* electrically programmable read only memory
erasable programmable read only memory...95, 98, 99, 100, 121, 125, 127
etching 88

F

farad 26
Faraday, Michael 26
ferromagnetic................................... 35
field programmable gate array.. 3, 114, 115, 116, 124, 127, 128, 129, 130, 131, 132, 133, 134
filter 38, 39, 40, 41
    band pass....................................... 41
    high pass.. 40
    low pass... 39

finite state machine. *See* state machine
Fisher, Josh 147
flash memory.................................. 100
Fleming, Ambrose............................ 47
flip-flop 70, 111, 126, 129, 130
FPGA .... *See* field programmable gate array
frequency...........38, 39, 40, 41, 72, 130
Frohman-Bentchkowsky, Dov........... 99
function block 125
fuse 97

H

Harvard architecture...................... 140
Hennessey, John 146
Henrich, William 97
henry... 29
Henry, Joseph 29
hertz... 32
Hertz, Heinrich................................. 31

I

I/O block ..125, 126, 128, 130, 131, 132
I/O device 140
IBM Corporation 101
ICE *See* in-circuit emulator
in-circuit emulator 161, 162
inductor ..23, 28, 29, 30, 38, 39, 40, 41, 57
instruction140, 141, 142, 143, 144, 147
insulator.....................47, 49, 50, 79, 87
integrated circuit ..59, 84, 85, 109, 112, 139, 161, *See* integrated circuit
Intel Corporation 84, 99, 104
interconnect 125, 127, 128, 131, 132
ion beam... 88
ion implantation 88

K

Kilby, Jack... 84

L

liquid resist 88
lithography 88
logic analyzer 159
logic probe 158

M

magnetism 6, 11, 26
mask 88, 96, 113
Masuoka, Fujio 100
microphone 35, 36, 39
microprocessor 114, 121, 139, 161
multiplexer 125, 129
mux .... *See* multiplexer
muxiplexer 126

N

nonlinear device 47, 49, 82
non-recurring expense ... 112, 113, 115, 116
nonvolatile random access memory 95, 98, 100, 101
Noyce, Robert 84
NRE .... *See* non-recurring expense
NVRAM *See* nonvolatile random access memory

O

ohm 24, 153, 154
Ohm, Georg 24
Ohm's Law 24, 27
ohmmeter 154
oscilloscope 155, 156
    analog 156
    digital 156
oxidation 86, 87
oxide layer 88

P

PAL .... *See* programmable array logic
Patterson, David 146
Perlegos, George 99
photolithography 88
photoresist 88
physical etching 88
piezoelectric 36
PLA .... *See* programmable logic array
printed circuit board 59, 60, *See* printed circuit board
processor 139, 140, 141, 142, 143, 144, 145, 146, 147
    accumulator 142, 143
    address register 141, 144
    arithmetic logic unit ... 142, 143, 144
    control logic 142, 143, 146
    control section 140
    data section 140
    decode stage 144
    execute stage 144
    general registers 141
    instruction fetch stage 143
    instruction register 142
    load/store stage 144
    pipelined 143, 144, 145
    program counter 141, 142
program branch 143
programmable array logic 123, 124, 125, 126, 127
Programmable device 121
programmable logic array 122, 123
programmable read only memory ... 95, 97, 98, 100, 101, 105, 121, 122, 127
PROM ...... *See* programmable read only memory
protocol analyzer 160

R

read only memory 95, 96, 139, 140
reduced instruction set computer . 146, 147
resistor 23, 24, 26, 27, 38, 39, 40, 41, 57, 61, 69, 96
RISC .... *See* reduced instruction set

Index

computer
ROM................... *See* read only memory

S

Salsbury, Phillip................................ 99
schematic.................................. 69, 101
semiconductor.... 79, 80, 81, 82, 83, 97,
   109, 110, 112, 133, 154
    n-type 81, 82
    p-type 81
Sequin, Carlo.................................... 146
Shannon, Claude............................... 65
Shockley, William.............................. 49
signal generator............................... 157
silicon dioxide 86, 87
silicon wafer....... 85, 86, 87, 88, 89, 114
speaker 35, 37
Stanford University 2, 3, 146
state machine 71, 72, 73, 109, 121, 124,
   129, 142
static random access memory . 95, 100,
   102, 104, 105, 132, 133, *See* static
   random access memory
switch matrix 127, 131

T

termination
    parallel.... *See* Thevenin termination
    serial 61
    Thevenin 61
Tesla, Nicola..................................... 31
Texas Instruments............................. 84
Toshiba Corporation 100
transducer 35, 36
transistor 47, 49, 50, 51, 69, 79, 83, 84,
   96, 97, 98, 99, 100, 101, 102, 104,
   109, 110, 111, 112, 113, 114, 131

base 49
bipolar junction transistor 49, 50
BJT .. *See* transistor, bipolar junction
collector... 49
drain 50
emitter.. 49
field effect transistor 50
gate... 50
source 2, 7, 8, 39, 50, 99
transistor, FET........ *See* transistor, field
   effect
transmission line............. 57, 58, 60, 61

U

University of California 146

V

very long instruction word.............. 147
VLIW ..... *See* very long instruction word
volt............. 9, 23, 25, 31, 32, 57, 58, 59
Volta, Alessandro................................ 8
voltmeter 153, 154
Von Neumann architecture 139
Von Neumann, John 139

W

Westinghouse, George 31

Y

Yale University 147

16. Answers to Quizzes

This section has the answers to all of the quizzes in the sections above.

Answers to Quizzes

Quiz 1: Electricity

1. **Electricity is**
   - the flow of electrons
2. **Which of the following have a positive electric charge?**
   - proton
   - nucleus (remember a nucleus consists of the positive protons and the neutral neutrons)
3. **Which of the following have a negative electric charge?**
   - electron
4. **Which of the following have no electric charge?**
   - neutron
5. **A chain reaction of electrons jumping from one atom to a nearby atom is called**
   - electricity
6. **What uses a chemical reaction to separate electrons from their atoms?**
   - electrochemical cells
   - voltaic cells (these are the different names for the same thing)
7. **A battery consists of one or more**
   - electrochemical cells
   - voltaic cells (these are the different names for the same thing)
8. **Batteries produce electricity by**
   - separated electrodes surrounded by an electrolyte

9. **Generators produce electricity by**
    - moving a coil of wire between the poles of a magnet
    - moving magnets around a coil of wire
10. **Which of the following types of generators do not heat water into steam**
    - active solar
11. **Throughout this book, to describe electric circuits I will use the analogy of**
    - water moving through pipes
12. **Albert Einstein was awarded his Nobel Prize for**
    - the photovoltaic effect

Quiz 2: Linear Devices

1. Match the component name in the box next to the corresponding symbol.

─/\/\/\─	Resistor
─┤├─	Capacitor
─⌒⌒⌒─	Inductor

2. Which of the following is not a basic component of electrical engineering?

    ■ gyrator

3. Components whose properties can be described using multiplication or division are called:

    ■ linear

4. The equation that states I=V/R is called:

    ■ Ohm's Law

5. A resistor can be thought of as a:

    ■ pipe with a cinch in it

6. Resistance is measured in:

    ■ ohms

Answers to Quizzes

7. **Current is measured in:**
   - amps
8. **Capacitance is measured in:**
   - farads
9. **A capacitor can be thought of as a:**
   - pipe with a bucket
10. **Inductance is measured in:**
    - henries
11. **An inductor can be thought of as a:**
    - pipe with a water wheel
12. **Hertz is a measure of:**
    - cycles per second

Quiz 3: Electricity as Sound

1. **An electromagnet is created by**
   - wrapping a wire around a core and passing an electric current through it

2. **The best core for an electromagnet is made from which material?**
   - ferromagnetic

3. **What does a transducer do?**
   - convert pressure into electricity

4. **What does a microphone uses a transducer to do?**
   - convert sound to electricity

5. **What does a speaker uses an electromagnet to do?**
   - convert electricity to sound

6. **The frequency of a signal describes**
   - how quickly the signal changes

7. **A low pass filter lets through**
   - low frequency signals

8. **A high pass filter lets through:**
   - high frequency signals

9. **A band pass filter lets through**
   - only signals in a certain range of frequencies

10. **The bass control on your stereo is**
    - a low pass filter

11. **The treble control on your stereo is**
    - a high pass filter

Answers to Quizzes

12. The cut-off frequencies of a filter are determined by

- all of the above (the value of the resistor, capacitor, and inductor)

Quiz 4: Nonlinear Devices

1. Enter the letter for the component name in the box next to the corresponding symbol.

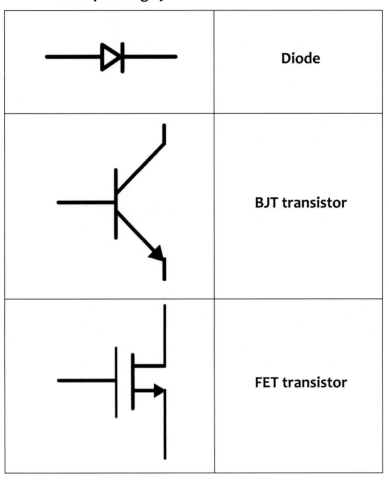

2. Components that cannot be defined by a simple multiplication or division equation are called

   ■ nonlinear

Answers to Quizzes

3. **Select the terminals of a diode**
   - cathode
   - anode

4. **A diode can be thought of as a**
   - pipe with a one-way valve

5. **A transistor can be thought of as a**
   - pipe with a valve that is controlled by another pipe

6. **Select the two basic kinds of transistors.**
   - field effect transistors
   - bipolar junction transistors

7. **Select the terminals of a BJT transistor**
   - collector
   - emitter
   - base

8. **In a BJT transistor, the base current controls the flow of electricity from**
   - the collector to the emitter

9. **Select the terminals of an FET transistor**
   - gate
   - drain
   - source

10. **In an FET transistor, the gate voltage controls the flow of electricity from**
    - the collector to the emitter

11. **The invention of transistors enabled**
    - all of the above (inexpensive radios, low power radios, modern computers, small, hand-held radios, efficient telephone circuits)
12. **Transistors replaced**
    - vacuum tubes

Answers to Quizzes

Quiz 5: Transmission Lines

1. **Transmission lines are any conductor that acts like a combination of many resistors, capacitors, and inductors.**
   - True

2. **All conductors are transmission lines.**
   - True

3. **Only high speed conductors act as transmission lines.**
   - False

4. **Reflection and absorption of electrical signals at the end of the transmission line causes electrical waves in the line.**
   - False

5. **A wave propagates on a printed circuit board (PCB) trace at the rate of:**
   - about 1 foot per 2 nanoseconds

6. **Transmission line effects on a PCB trace cause smoother, more reliable signals.**
   - False

7. **Critical PCB traces should be, as much as possible, straight lines.**
   - True

8. **Critical PCB traces should be as long as possible.**
   - False

9. **The receivers on critical PCB traces should be spread apart.**
   - False

10. **Serial termination involves**
    - putting resistors between the drivers and the trace

11. **Parallel or Thevenin termination involves**
    - putting resistors at the end of the trace, connected to power and ground
12. **Power lines to your house are transmission lines.**
    - True

Quiz 6: Digital Logic

1. **Given the following three statements and their logic values, determine whether the next statements are true or false according to Boolean logic.**

    A. The sun is shining (TRUE)

    B. This earth is flat (TRUE)

    C. Everybody cares about science (FALSE)

    - FALSE   The earth is flat and the sun is not shining.
    - FALSE   The earth is not flat or the sun is not shining.
    - TRUE   The earth is flat and not everybody cares about science.
    - TRUE   Everybody cares about science or the sun is shining.

2. **Given the following three logic values, determine whether the next Boolean logic statements are true or false.**

    A. TRUE

    B. TRUE

    C. FALSE

    - FALSE   B AND NOT A
    - FALSE   NOT B OR NOT A
    - TRUE   B AND NOT C
    - TRUE   C OR A

Answers to Quizzes

3. Given the following three Boolean values, determine whether the next Boolean equations equal 1 or 0.

    A. 1

    B. 1

    C. 0

    B & ~A = 0

    ~B | ~A = 0

    B & ~C = 1

    C | A = 1

4. Which of the following are basic Boolean operators?
    - NAND
    - OR
    - NOR
    - AND
    - NOT
    - XOR
    - XNOR

5. What is the number base for decimal numbers?
    - 10

6. What is the number base for binary numbers?
    - 2

7. Delay elements are implemented in digital logic using
    - flip-flops

Just Enough Electronics to Impress Your Friends and Colleagues

8. **Match the schematic symbol with its Boolean function.**

	NOR
	XNOR
	NAND
	AND
	OR
	NOT
	XOR

Answers to Quizzes

9. Fill in the missing numbers from the binary counting sequence.

000	0
001	1
010	2
<u>011</u>	3
100	4
101	5
<u>110</u>	6
<u>111</u>	7

10. Calculate the answer to the following addition problem in binary

```
   010          2
 + 011        + 3
   ---        ---
   101          5
```

11. Calculate the answer to the following multiplication problem in binary

```
   010          2
 x 011        x 3
   ---        ---
   110          6
```

12. Finite state machine
    - all of the above (combinations of Boolean logic and clocked elements, typically implemented with Boolean logic gates flip-flops, can implement any form of algorithm, often designed using state diagrams)

Quiz 7: Semiconductors

1. Match the term to its correct definition.

a) Conductor	Allows a large amount of electricity to flow through it when a voltage is applied.
b) Insulator	Does not allow a significant amount of electricity to flow through it when a voltage is applied.
c) Semiconductor	Under some conditions, does not allow much electricity to flow though it. Under other conditions, allows a large amount of electricity to flow through it.

2. Check all three basic parts of an atom.

    ■ electron

    ■ proton

    ■ neutron

3. A material with a very regular atomic structure, where the atoms are all lined up, is called

    ■ a crystal

4. The most common semiconductor material used to create integrated circuits is

    ■ silicon

5. When we "dope" a silicon crystal with atoms that have more free electrons than the silicon does, the resulting substance is called

    ■ n-type semiconductor

Answers to Quizzes

6. When we "dope" a silicon crystal with atoms that have fewer free electrons than the silicon does, the resulting substance is called

    ■ p-type semiconductor

7. A junction of p-type semiconductor material and n-type semiconductor material is called a pn junction, or a

    ■ diode

8. A sandwich of n-type, p-type, and n-type semiconductor materials is called a

    ■ transistor

9. A semiconductor device that is made up of many transistors is called

    ■ an integrated circuit

10. A tube of silicon drawn out from a vat of liquid silicon and cooled is called

    ■ a boule

11. A round, polished slice of silicon is called

    ■ a wafer

12. Which one of the following is not a step in the process of creating silicon wafers?

    ■ residualization

Quiz 8: Memories

1. Which of the following memories lose data when they are powered down? (check all that apply)
   - SRAM
   - DRAM

2. Which of the following memories can have its data overwritten? (check all that apply)
   - EPROM
   - EEPROM
   - NVRAM
   - SRAM
   - DRAM

3. Which of the following memories cannot be erased? (check all that apply)
   - ROM
   - PROM

4. A ROM bit is programmed with (check all possibilities)
   - a metal connection and a transistor

5. A PROM bit is programmed with (check all possibilities)
   - a transistor and a fuse
   - a diode and an antifuse

6. An EPROM bit is programmed with (check all possibilities)
   - electric charge on the input of a transistor

Answers to Quizzes

7. **An NVRAM can be a combination of (check all possibilities)**
   - DRAM and EEPROM
   - SRAM and battery
   - SRAM and EEPROM

8. **Which of the following memories generally has the highest density of bits?**
   - DRAM

9. **Which of the following memories needs to be regularly refreshed?**
   - DRAM

10. **Which of the following memories is generally fastest?**
    - SRAM

11. **A 16 by 4 memory has 4 address lines and 4 data lines.**

12. **A 128 byte memory has 7 address lines and 8 data lines.**

Quiz 9: ASICs

1. **What does the term ASIC stand for?**

   ■ Application Specific Integrated Circuit

2. **Give the name of the ASIC architecture shown below.**

   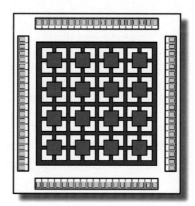

   ■ Structured ASIC

3. **Which of the following things does an ASIC vendor not supply?**

   ■ The high-level chip design

4. **Which of the following things does an ASIC designer need to know?**

   ■ High-level chip design

5. Give the name of the ASIC architecture shown below.

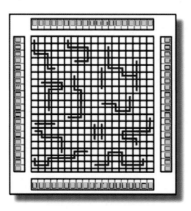

- Gate Array

6. Which of the following ASIC architectures consists of rows and columns of large cells containing logic and registers?

- Structured ASIC

7. Which of the following ASIC architectures consists of rows and columns of regular transistor structures?

- Gate array

8. Give the name of the ASIC architecture shown below.

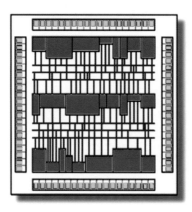

- Standard Cell

9. Which of the following ASIC architectures is designed using cells of transistors that are already connected together and compactly routed to form higher level functions such as flip-flops, adders, counters, multipliers, and even entire processors?

   ■ Standard cell

10. Which of the following ASIC architectures is the only one that has so far survived in the market against Field Programmable Gate Array (FPGA) technology?

    ■ Standard cell

11. Select all of the advantages for each ASIC architecture in the table below.

	Gate Array		Standard Cell		Structured ASIC	
Initial cost (NRE)	a.	low	b.	high	c.	low
Per piece cost	d.	high	e.	low	f.	high
Utilization	g.	low	h.	high	i.	low
Turn around time	j.	fast	k.	slow	l.	fast

12. Which of the following ASIC architectures is most similar to the architecture of a Field Programmable Gate Array (FPGA)?

    ■ Structured ASIC

Answers to Quizzes

Quiz 10: Programmable Devices

1. **What does the term PROM mean?**
   - Programmable Read Only Memory

2. **What does the term PLA mean?**
   - Programmable Logic Array

3. **A PLA contains (check all that apply)**
   - A large programmable AND plane
   - A large programmable OR plane
   - Inverters on the inputs and outputs

4. **What does the term PAL mean?**
   - Programmable Array Logic

5. **A PAL contains (check all that apply)**
   - A large programmable AND plane
   - Inverters on the inputs and outputs
   - A small fixed OR plane

6. **What does the term CPLD mean?**
   - Complex Programmable Logic Device

7. **A CPLD contains (check all that apply)**
   - A single large programmable switch matrix
   - I/O blocks
   - Programmable elements
   - Clock buffers and clock lines
   - Function blocks like individual PALs

8. **What does the term FPGA mean?**
   - Field Programmable Gate Array

Answers to Quizzes

9. An FPGA contains (check all that apply)
   - Combinatorial logic blocks containing small RAMs, flip-flops, and muxes
   - I/O blocks
   - Programmable elements
   - Clock buffers and clock lines
   - A hierarchy of interconnect lines

10. Which technology is not used in a CPLD for programming?
    - SRAM

11. Which technology is not used in a FPGA for programming?
    - DRAM

12. Match each programmable device with its description.

a) PROM	A memory device that can be programmed once and read many times.
b) PLA	A logic device with a large AND plane and a large OR plane for implementing different combinations of Boolean logic.
c) PAL	A logic device with a large AND plane and a small, fixed number of OR gates for implementing Boolean logic and state machines.
d) CPLD	A logic device that is made up of many PAL devices.
e) FPGA	A logic device that can be used to design large functions like an ASIC except that it can be programmed quickly and inexpensively.

Quiz 11: Processors

1. The man who invented the modern computer architecture was?
   - John Von Neumann

2. The main parts of a general purpose computer are (check all that apply)
   - ROM
   - processor
   - RAM
   - I/O devices

3. Which are not buses in a general purpose computer architecture?
   - school bus
   - control bus
   - universal serial bus
   - Harvard bus

4. The two main sections of a computer processor are
   - data section
   - control section

5. Check all of the functions that can be found in the data section of a computer processor.
   - ALU
   - address register
   - accumulator
   - program counter
   - general registers

Answers to Quizzes

6. Check all of the functions below that can be found in the control section of a computer processor.
   - control logic
   - instruction register

7. What does ALU mean?
   - Arithmetic Logic Unit

8. Select all of the following that can be stages of a pipelined processor.
   - load/store
   - decode
   - execute
   - instruction fetch

9. The small, fast memory that is used to speed up accesses to the main memory is called a
   - cache memory

10. What does the term RISC mean?
    - reduced instruction set computer

11. What does the term CISC mean?
    - complex instruction set computer

12. What does the term VLIW mean?
    - very long instruction word

Quiz 12: Equipment

1. Which three of the following devices might be found in a digital multimeter?
   - voltmeter
   - ammeter
   - ohmmeter

2. What does a voltmeter measure?
   - voltage

3. What does an ammeter measure?
   - current

4. What does an ohmmeter measure?
   - resistance

5. The electrocardiogram (ECG) heart monitor in hospitals, which shows a line for the patient's heartbeat, is essentially
   - an oscilloscope

6. The term "ICE" stands for
   - in-circuit emulator

Answers to Quizzes

7. **Match the equipment to its function.**

a) Oscilloscope	Used to look at voltage signals at points on a circuit, as the signals change in time.
b) Signal generator	Used to produce very specific types of signals that are input to a circuit for testing.
c) Logic probe	A pen-like device that is used to examine digital systems.
d) Logic analyzer	Used to examine complex digital signals by storing the signal levels at specific times.
e) Protocol analyzer	Stores digital signals that are transmitted through it, decodes them, and interprets them.
f) In-circuit emulator	Used to replace a microprocessor in a real circuit for testing the circuit.